Battery Technology for Electric Vehicles

Electric drive vehicles (EDVs) are seen on American roads in increasing numbers. Related to this market trend and critical for it to increase are improvements in battery technology. *Battery Technology for Electric Vehicles* examines in detail the research support from the U.S. Department of Energy (DOE) for the development of nickel metal hydride (NiMH) and lithium-ion (Li-ion) batteries used in EDVs. With public support comes accountability of the social outcomes associated with public investments.

The book overviews DOE investments in advanced battery technology, documents the adoption of these batteries in EDVs on the road, and calculates the economic benefits associated with these improved technologies. It provides a detailed global evaluation of the net social benefits associated with DOE investments, the results of the benefit-to-cost ratio of over 3.6-to-1, and the life-cycle approach that allows adopted EDVs to remain on the road over their expected future life, thus generating economic and environmental health benefits into the future.

Albert N. Link is Professor of Economics at the University of North Carolina at Greensboro, USA. His research is related to the economics of innovation, technology policy, and program evaluation.

Alan C. O'Connor is an economist and Director of Innovation Economics at RTI International. He specializes in economic analysis of research and development (R&D) programs, program evaluation, and economic development.

Troy J. Scott is an economist at RTI International, where his research deals with the economics of technology and innovation. His work focuses on the nexus of public support for research and development (R&D), regulation, and R&D rivalry among firms to evaluate and inform public policy.

"Fifty years ago Edwin Mansfield used economics and econometrics with in-depth case studies to transform our understanding of innovation. Since 1972, federal agencies have invested over a billion dollars in the battery technologies important to electric vehicles. Link, O'Connor, and Scott use the 'Mansfield' strategy to take readers 'under the hood' and ask if these programs were in the public interest. Their book is a great read!"

V. Kerry Smith, Arizona State University, USA

"The authors address an important issue which is high on the policy agenda in many industrialized countries. Even using conservative estimates about social benefits of public support for new technologies, they find substantial ones. In the vein of discussing public/private partnerships in science and technology, this study is a must-read for policy makers and research funders in the field."

Wolfgang Polt, Institute for Economic and Innovation Research, Austria

"This tome presents a thorough empirical economic evaluation of the social benefits attributable to federal R&D investment in vehicle battery technology in the United States. Link, O'Connor, and Scott have produced one of the best such appraisals available. A must-read."

Nicholas S. Vonortas, George Washington University, USA

Battery Technology for Electric Vehicles

Public science and private innovation

Albert N. Link, Alan C. O'Connor, and Troy J. Scott

Routledge
Taylor & Francis Group
LONDON AND NEW YORK

earthscan
from Routledge

First published 2015
by Routledge
2 Park Square, Milton Park, Abingdon, Oxon OX14 4RN

and by Routledge
711 Third Avenue, New York, NY 10017

Routledge is an imprint of the Taylor & Francis Group, an informa business

British Library Cataloguing-in-Publication Data
A catalogue record for this book is available from the British Library

Library of Congress Cataloging-in-Publication Data
Link, Albert N.
Battery technology for electric vehicles : public science and private innovation / Albert N. Link, Alan C. O'Connor, and Troy J. Scott.
pages cm
Includes bibliographical references and index.
1. Electric vehicles – Batteries. 2. Electric vehicles – Cost effectiveness.
I. O'Connor, Alan C. II. Scott, Troy J. III. Title.
TL220.L56 2015
629.25 – dc23
2014039612

ISBN: 978–1–138–81110–2 (hbk)
ISBN: 978–1–315–74930–3 (ebk)

Typeset in Goudy
by Swales & Willis Ltd, Exeter, Devon, UK

Printed and bound in the United States of America by Publishers Graphics, LLC on sustainably sourced paper.

For Carol, Scott, and Dorothy

Contents

Figures

Tables

Acknowledgements

A number of individuals contributed to this study, both in terms of their participation during the data collection effort and in terms of their comments and suggestions on earlier versions of these chapters.

We are grateful to the scientists, engineers, and analysts that contributed data and insight that informed the evaluation portion of this study, including those from A123 Systems (Navitas Systems), Amprius, Applied Materials, BASF Materials USA, BASF Catalysts, Dow Kokam, EnerDel, FMC Corporation, Ford Motor Company, General Motors, H&T Waterbury, K2 Energy Solutions, LG Chem Power, Maxwell Technologies, Miltec UV International, Nanosys, Saft America, Seeo, Ultimate Membrane Technologies, Northwestern University, Pennsylvania State University, SUNY Binghamton, University of Massachusetts Boston, University of Pittsburgh, University of Rhode Island, Argonne National Laboratory, Brookhaven National Laboratory, Idaho National Laboratory, Lawrence Berkeley National Laboratory (LBNL), National Renewable Energy Laboratory, Oak Ridge National Laboratory, Pacific Northwest National Laboratory, Southwest Research Institute, and U.S. Army Research Laboratory.

We are also grateful to a number of individuals who offered valuable comments and suggestions (alphabetically): Tien Duong (Vehicle Technologies Office [VTO] in the Office of Energy Efficiency and Renewable Energy [EERE] within the Department of Energy [DOE]), Irwin Feller (consultant to the American Association for the Advancement of Science), David Finifter (The College of William & Mary), Michael Gallaher (RTI International), Ken Keating (consultant to EERE), David Howell (VTO), William Key (VTO), Cheryl Oros (Oros Consulting), Rosalie Ruegg (TIA Consulting), Edward Vine (LBNL), and Thomas White (DOE Office of Policy).

A special thanks to Yaw Agyeman (LBNL) and Jeff Dowd (EERE) who supported financial assistance to conduct this study. And, a special thanks to the Center on Globalization, Governance and Competitiveness at Duke University for the use of Figure 4.1.

Lastly, we are indebted to Sara Casey, Ross Loomis, and Lynn Davis, all of RTI International, for their invaluable contributions to this project.

Abbreviations

ACE	advanced combustion engine
AEC	Atomic Energy Commission
ARRA	American Recovery and Reinvestment Act
a-Si	amorphous silicon
BCR	benefit-to-cost ratio
BEA	Bureau of Economic Analysis
CAFE	Corporate Average Fuel Economy
CARB	California Air Resources Board
CDC	Centers for Disease Control and Prevention
CdTe	cadmium telluride
CFC	chlorofluorocarbon
CIS	copper indium diselenide
CO_2eq	equivalent CO_2
COBRA	Co-Benefits Risk Assessment
CRADA	Cooperative Research and Development Agreement
CRF	Combustion Research Facility
c-Si	crystalline silicon
DOD	depth of discharge
DOE	Department of Energy
ECD	Energy Conversion Devices
EDV	electric drive vehicle
EERE	Office of Energy Efficiency and Renewable Energy
EPA	Environmental Protection Agency
EPAct	Energy Policy Act of 1992
EPRI	Electric Power Research Institute
ERDA	Energy Research and Development Administration
EV	electric vehicle
FSA	Flat-Plate Solar Array Project
GHG	greenhouse gas
GM	General Motors
GTP	Geothermal Technologies Program
GWP	global warming potential

HEV	hybrid electric vehicle
ICV	internal-combustion vehicle
IDMS	Isotope Dilution Mass Spectrometry
IRR	internal rate of return
JCS	Johnson Controls/Saft
JPL	Jet Propulsion Laboratory
LAS	laser absorption spectrometry
LDV	Laser Doppler velocimetry
LIF	laser-induced fluorescence
LII	laser-induced incandescence
Li-ion	Lithium-ion
LRS	laser Raman spectroscopy
MATS	Mercury and Air Toxics Standards
mpg	miles per gallon
MRAD	minor restricted activity days
MW	megawatt
NiMH	nickel metal hydride
NIST	National Institute of Standards and Technology
NO_x	nitrogen oxides
NPV	net present value
NRC	National Research Council
NREL	National Renewable Energy Laboratory
OEM	original equipment manufacturers
OMB	Office of Management and Budget
OPEC	Organization of the Petroleum Exporting Countries
OTA	Office of Technology Assessment
OTP	Office of Technology Policy
OTT	Office of Transportation Technologies
PDC	polycrystalline diamond compact
PHEV	plug-in hybrid electric vehicles
PIV	particle image velocimetry
PM	particulate matter
PNGV	Partnership for a New Generation of Vehicles
PtW	pump-to-wheel
PV	photovoltaic
PVMaT	PV Manufacturing Technology Project
R&D	research and development
SCC	Social Cost of Carbon
SRM	Standard Reference Material
SETP	Solar Energy Technology Program
TCP	Thermocouple Calibration Program
TFPs	Thin-Film PV Partnerships
TOUGH	Transport of Unsaturated Groundwater and Heat
USABC	U.S. Advanced Battery Consortium
USCAR	U.S. Council for Automotive Research

VOCs	volatile organic compounds
VTO	Vehicle Technologies Office
VTP	Vehicle Technologies Program
WtP	well-to-pump
WtW	well-to-wheel

1 Introduction

The policy imperative

At the time of the Organization of the Petroleum Exporting Countries (OPEC) oil embargo, the only U.S. government agency related to energy was the Atomic Energy Commission (AEC).[1,2] In response to the OPEC oil embargo, President Nixon launched Project Independence on November 7, 1973; the goal of the project was to achieve energy independence by 1980. In his State of the Union Address on January 30, 1974, President Nixon remarked (Nixon, 1974):

> Let it be our national goal: At the end of this decade, in the year 1980, the United States will not be dependent on any other country for the energy we need to provide our jobs, to heat our homes, and to keep our transportation moving.

Others at that time also editorialized about the importance of the oil embargo on the future direction of U.S. energy policy (Dooley 2008, p. 9):

> The [OPEC] Oil Embargo which began on October 19, 1973 sparked a fundamental reassessment of the nation's vulnerability to imported energy and also forced a reassessment of the role that energy R&D could play in helping secure the nation against hostile acts like the Oil Embargo.

The United States' heightened interest in alternative energy sources led in 1975 to replacement of the AEC by the Energy Research and Development Administration (ERDA) in an effort to unify the federal government's energy R&D activities. Congress charged ERDA to sponsor research and development (R&D) related to electric and hybrid vehicles through the passage of the Electric and Hybrid Vehicle Research, Development, and Demonstration Act of 1976, Public Law 94-413. Therein:[3]

> The Congress finds and declares that:
>
> 1 the Nation's dependence on foreign sources of petroleum must be reduced, as such dependence jeopardizes national security, inhibits foreign policy, and undermines economic well-being;

2 the Nation's balance of payments is threatened by the need to import oil for the production of liquid fuel for gasoline-powered vehicles;
3 the single largest use of petroleum supplies is in the field of transportation, for gasoline- and diesel-powered motor vehicles;
4 the expeditious introduction of electric and hybrid vehicles into the Nation's transportation fleet would substantially reduce such use and dependence.

On August 4, 1977, President Carter signed the Department of Energy Reorganization Act of 1977, Public Law 95-91, transferring the mission of ERDA to the newly formed Department of Energy (DOE). As stated in the Act, Congress finds that:

- the United States faces an increasing shortage of nonrenewable energy resources;
- this energy shortage and our increasing dependence on foreign energy supplies present a serious threat to the national security of the United States and to the health, safety and welfare of its citizens;
- a strong national energy program is needed to meet the present and future energy needs of the Nation consistent with overall national economic, environmental and social goals;
- responsibility for energy policy, regulation, and research, development and demonstration is fragmented in many departments and agencies and thus does not allow for the comprehensive, centralized focus necessary for effective coordination of energy supply and conservation programs; and
- formulation and implementation of a national energy program require the integration of major Federal energy functions into a single department in the executive branch.

By this act, Congress declared that the establishment of a Department of Energy in the Executive Branch is in the public interest and will promote the general welfare by assuring coordinated and effective administration of Federal energy policies and programs. DOE will:

> carry out the planning, coordination, support, and management of a balanced and comprehensive energy research and development program, including – (A) assessing the requirements for energy research and development; (B) developing priorities necessary to meet those requirements; (C) undertaking programs for the optimal development of the various forms of energy production and conservation; and (D) disseminating information resulting from such programs.

Motivated by the Electric and Hybrid Vehicle Research, Development, and Demonstration Act of 1976, and the subsequent availability of public funding,

Chrysler (now Chrysler Group LLC), Ford Motor Company, and General Motors (GM) established in early 1991 the U.S. Advanced Battery Consortium (USABC) to accelerate the development of batteries for electric drive vehicles (EDVs). The term 'EDV' refers to all types of electric drive vehicles including:[4]

- hybrid electric vehicles (HEVs), which use gasoline to charge the battery and part of the time to power the vehicle (e.g., the first-generation Prius);
- electric vehicles (EVs), which are powered exclusively by a battery and must be plugged into an electrical outlet to recharge (e.g., the Tesla Model S, Nissan Leaf); and
- plug-in hybrid electric vehicles (PHEV), which can either use gasoline to recharge the battery and power the vehicle or be plugged in to recharge the battery (e.g., the Chevrolet Volt).

The creation of the USABC was also motivated, in part, by the recent California Air Resources Board's (CARB's) 1990 regulations for low-emission vehicles and its clean fuel standards for emissions that were to be applied to new classes of vehicles not later than 1994. USABC's purpose was to:

> work with advanced battery developers and companies that will conduct research and development (R&D) on advanced batteries to provide increased range and improved performance for electric vehicles in the latter part of the 1990s.
>
> (National Research Council (NRC) 1998, p. 12)

More specifically, the USABC had the following overarching objectives:

> - to establish a capability for an advanced battery manufacturing industry in the United States;
> - to accelerate the market potential of EVs through joint research on the most promising advanced battery alternatives;
> - to develop electrical energy systems capable of providing EVs with ranges and performance levels competitive with petroleum-based vehicles;
> - to leverage external funding for high-risk, high-cost R&D on advanced batteries for EVs.
>
> (NRC 1998, p. 21)

DOE joined the consortium in late 1991 in response to its mandate through the Electric and Hybrid Vehicle Research, Development, and Demonstration Act of 1976. And, this mandate was reconfirmed through the Energy Policy Act of 1992 (EPAct).[5] In addition, the Electric Power Research Institute (EPRI) joined the consortium in 1991.[6]

Related to the ongoing charge for DOE's involvement in electric and hybrid vehicles and related battery research, President Clinton initiated the Partnership for a New Generation of Vehicles (PNGV) program in 1993.

This was a cooperative R&D program between the federal government and the U.S. Council for Automotive Research (USCAR), which included Chrysler, Ford, GM, and relevant federal agencies and the national laboratories (Sissine 1996).[7] Noteworthy was one of the original technology goals of PNGV (Sissine 1996, p.1):

> Research and development goals for industry and government engineering teams have been launched in three categories: advanced manufacturing techniques that help get new product ideas more quickly into the marketplace; technologies that can lead to near-term improvements in automobile efficiency, safety, and emissions; and research that could lead to production prototypes of vehicles capable of up to 80 miles per gallon – three times greater fuel efficiency than the average car of today.

More specifically, the goals of the PNGV were (NRC 2001, p. 146):

> (1) to improve national manufacturing competitiveness, (2) to implement commercially viable technologies that increase the fuel efficiency and reduce the emissions from conventional vehicles, and (3) to develop technologies for a new class of vehicles with up to three times the fuel efficiency of 1994 midsize family sedans (80 mpg) while meeting emission standards and without sacrificing performance, affordability, utility, safety, or comfort.

A more fuel-efficient car might achieve the stated goal of 80 miles per gallon (mpg). But, a 1995 Office of Technology Assessment (OTA) report stated that there was at that time (i.e., 1993) no battery technology capable of achieving the equivalent of 80 mpg. However, the report went on to state that: "Nickel metal-hydride batteries are seen as the only longer-term battery technology that could possibly be designed to reach the 80 mpg target" (OTA 1995, p. 17).[8]

Overview of EERE R&D support for battery technology

Within DOE, the Office of Energy Efficiency and Renewable Energy (EERE) "accelerates development and facilitates deployment of energy efficiency and renewable energy technologies and market-based solutions that strengthen U.S. energy security, environmental quality, and economic vitality."[9] EERE leads DOE's "efforts to develop and deliver market-driven solutions for energy-saving homes, buildings, and manufacturing; sustainable transportation; and renewable electricity generation."[10]

EERE consists of several offices and programs that support its mission.[11] Related to energy efficiency are the Advanced Manufacturing Office, the Buildings Technology Office, the Federal Energy Management Program, the Weatherization and Intergovernmental Program, and the Sustainability Performance Office. Related to renewable power are the Geothermal Technologies Office, the Solar Energy Technologies Office, the Wind Program, and the Water Power

Program. And regarding transportation are the Bioenergy Technologies Office, the Hydrogen and Fuel Cell Technologies Office, and the Vehicle Technologies Office (VTO).

The mission of VTO – the R&D program that is the focus in this book – is:[12]

> to develop more energy efficient and environmentally friendly highway transportation technologies that enable America to use less petroleum. The long-term aim is to develop "leap frog" technologies that will provide Americans with greater freedom of mobility and energy security, with lower costs and lower impacts on the environment.

Energy storage technology development is an essential element of VTO's mission:[13]

- Energy storage technologies, especially batteries, are critical enabling technologies for developing advanced, fuel-efficient, light- and heavy-duty vehicles, which are key components of DOE's Energy Strategic Goal: "to protect our national and economic security by promoting a diverse supply and delivery of reliable, affordable, and environmentally sound energy."
- VTO "supports the development of durable and affordable advanced batteries that cover the full range of vehicle applications, from start/stop to full-power hybrid electric, electric, and fuel cell vehicles."[14]
- Energy storage research aims to overcome specific technical barriers that have been identified by the automotive industry together with VTO – cost, performance, life, and abuse tolerance. These barriers are being addressed collaboratively by DOE's technical research teams and battery manufacturers.

DOE has invested in energy storage technologies since 1976. Figure 1.1 shows VTO's R&D investments in energy storage technologies from 1976 through 2012 in nominal and in real 2012 dollars (2012$). The R&D data that underlie Figure 1.1 are in Table 1.1.

VTO's funding toward advanced battery research in nickel metal hydride (NiMH) and lithium-ion (Li-ion) battery technology, in particular, began in 1992. More specifically, VTO's R&D investments in energy storage technology totaled $1,168 million in 2012 dollars from 1976 through 2012 (see Figure 1.1). The $197 million invested prior to 1992 supported general energy storage technologies such as batteries other than NiMH and Li-ion, flywheels, and ultracapacitors; as well as testing methods and standards development. The $971 million invested from 1992 through 2012 included primarily VTO's support for NiMH and Li-ion battery technologies.

The remainder of this book presents the results of an economic evaluation of the net social benefits attributable to VTOs R&D investments in battery technologies – NiMH and Li-ion battery technologies in particular. The premise

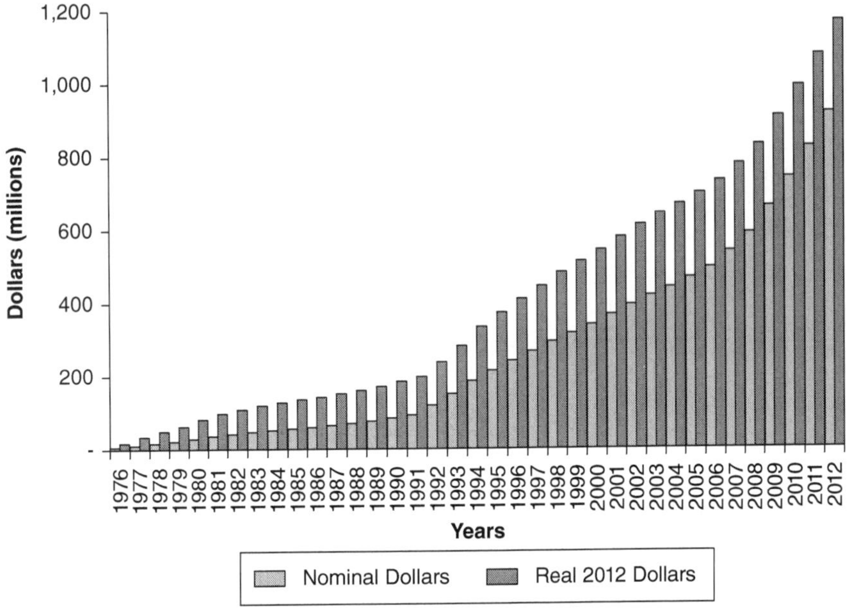

Figure 1.1 Cumulative VTO R&D investments in energy storage technologies, 1976 through 2012

Sources: Investment data provided by DOE. GDP chain-type price index from U.S. Department of Commerce, Bureau of Economic Analysis, downloaded from the St. Louis Federal Reserve, http://research.stlouisfed.org/fred2/series/GDPCTPI/downloaddata?cid=21.

Note: The data underlying Figure 1.1 are presented in Table 1.1.

of the evaluation is that these investments accelerated the development of battery technology relative to the timeline of development that would have unfolded in a counterfactual scenario without VTO investments. Central to this premise is the notion that public R&D investments may have positive social net present value although the same investments would not have been undertaken by the private sector alone. The economic arguments for why private underinvestment in R&D may be expected and how public-sector involvement may then improve efficiency are discussed in Chapter 2. The extent to which public investments did in fact accelerate technological development was ascertained through interviews with 54 experts in vehicle energy storage technologies, representing VTO-funded battery companies, car companies, research laboratories, and universities. With respect to NiMH technology, one interviewee noted that "without DOE, there would be essentially no U.S. [energy storage] industry. Technology would still have been developed abroad in, for example, Japan and Korea, and EDVs would still have made their way into the U.S. market, but it would have taken

Table 1.1 VTO's R&D investments in battery technologies, 1976 through 2012

Fiscal year	(1) Nominal appropriations (thousands of current $s)	(2) Price index (2005 = 100)	(3) Price index (2012 = 100)	(4) Appropriations (thousands of 2012$s)
1976	5,300	35.5	30.8	17,209
1977	5,300	37.8	32.8	16,180
1978	5,500	40.4	35.1	15,690
1979	5,200	43.8	38.0	13,695
1980	7,700	47.8	41.4	18,580
1981	7,100	52.3	45.4	15,656
1982	5,300	55.5	48.1	11,015
1983	5,500	57.7	50.0	10,997
1984	4,400	59.9	51.9	8,479
1985	4,280	61.7	53.5	8,006
1986	3,622	63.1	54.7	6,628
1987	5,220	64.8	56.2	9,292
1988	5,174	67.0	58.1	8,904
1989	6,417	69.6	60.3	10,641
1990	7,870	72.3	62.6	12,564
1991	8,836	74.8	64.9	13,625
1992	26,412	76.6	66.4	39,783
1993	30,911	78.3	67.9	45,557
1994	35,816	79.9	69.3	51,699
1995	27,724	81.6	70.7	39,200
1996	26,770	83.2	72.1	37,145
1997	25,497	84.6	73.3	34,763
1998	27,738	85.6	74.2	37,397
1999	22,784	86.8	75.3	30,272
2000	23,433	88.7	76.9	30,474
2001	27,706	90.7	78.6	35,234
2002	26,667	92.2	79.9	33,375
2003	24,517	94.1	81.6	30,051
2004	22,637	96.8	83.9	26,987
2005	25,855	100.0	86.7	29,832
2006	28,134	103.2	89.5	31,444
2007	44,412	106.2	92.1	48,238
2008	49,048	108.6	94.1	52,128
2009	73,425	109.5	94.9	77,347
2010	78,921	111.0	96.2	82,035
2011	83,299	113.4	98.3	84,778
2012	93,034	115.4	100.0	93,034

Sources: Appropriations data provided by DOE. GDP chain-type price index from U.S. Department of Commerce, Bureau of Economic Analysis, downloaded from the St. Louis Federal Reserve, http://research.stlouisfed.org/fred2/series/GDPCTPI/downloaddata?cid=21.

Note: Nominal appropriations include battery storage R&D and Small Business Innovation Research (SBIR) funded R&D. FY 1998 funding is an estimate of Phase I awards made in years prior to 1999. The estimate is based on Phase II awards between 1999 and 2003 for which no Phase 1 awards are tabulated. SBIR funding is the total funds awarded in a given year for automotive-related energy storage projects. The awards are made as a result of the solicitation and selection process managed by the Office of Science within the U.S. Department of Energy.

longer." Another interviewee noted that VTO's impact on Li-ion technology was still greater: "It is possible that without the [VTO's] support for battery technology development, there might presently be no Li-ion technology in the EDV market." These comments are representative of the connection, drawn by interviewees, between advancements in the state-of-the-art battery technology and the diffusion of that technology through the market adoption of EDVs: Only as the range and performance of EDVs improved and as costs were lowered did EDVs become a viable alternative to conventional internal-combustion vehicles (ICVs). The benefits quantified in this evaluation – and used to determine the social rate of return on VTO's investments – stem from the accelerated diffusion of EDVs that can be attributed to VTO. Specifically, each EDV on the U.S. market is assumed to have displaced an ICV, with a corresponding reduction in the consumption of fossil fuel; for a fraction of those EDVs (its estimated size based on the perceptions and opinions of interviewees) the benefits associated with this reduced fossil fuel consumption are attributed to VTO. The remaining sections of this chapter describe the evaluation approach in greater detail.

An overview of the evaluation approach

The evaluation herein builds on the pioneering work of Griliches (1958) and Mansfield et al. (1977) to develop estimates of the social rate of return to VTO's investments in battery R&D.[15] The Griliches/Mansfield methodology holds fixed, in the counterfactual situation, the status-quo technology that prevailed prior to the new technology brought forth with support from public R&D investments. Our approach is a modified version of this one. We recognize that the status-quo technology would not have remained fixed in the absence of VTO investments. Rather, the counterfactual situation involves battery technology being developed and adopted (through the sale of EDVs) at a different rate. We estimated counterfactual rates of development and adoption based on our interviews and derived the stream of estimated benefits from the comparison of this counterfactual description with the actual observed time series of EDV adoption. Streams of VTO investment outlays (the costs) are compared with estimates of the streams of economic surplus those investments have generated (the benefits), by means of conventional evaluation metrics: net present values, benefit-to-cost ratios, and internal rates of return.[16]

Social benefits beginning in 1999 and continuing through 2022 are quantified, and they are compared with VTO's R&D investments from 1992 through 2012.[17] The timeline of the stream of benefits begins in 1999 with the first EDVs on the road in the United States. EDVs are assumed to be driven for 11 years (with details of annual mileage assumptions provided in Chapter 4); the stream of benefits therefore ends in 2022, the last year in which EDVs purchased in 2012 are driven. A conservative, lower-bound estimate of benefits is also considered, truncating the stream of benefits at 2012.

Three categories of social benefits are considered: economic and energy benefits, environmental health benefits, and energy security benefits.[18] The principal source of EDVs' energy and resource benefits is EDVs' ability to operate under electric power for some or all of the time. Whereas the average ICV's fuel economy is 23.5 mpg, the equivalent is 34.8 mpg for hybrids, 40.8 mpg equivalent for plug-in hybrids, and 82.3 mpg equivalent for electric vehicles (see Huo et al. 2009). Under the assumption that the miles driven in EDVs displace the same miles driven by an ICV, a number of attributable EDV miles driven can be translated into attributable reductions in fuel consumption. This reduced fuel consumption has social value, equal to the full social cost of the fuel had it been consumed. Of this social cost, the market price of gasoline is a reasonable lower bound estimate, and this estimate can be improved by adding to it the value of co-benefits: the avoided costs associated with the health impacts of automobile exhaust.

Economic and energy benefits are related to the value of goods and services in the economy. Advancements in technology can increase the flow of economic benefits, through both improvements in the performance of existing goods and services and reductions in the cost of producing existing goods and services. Resource savings – such as energy savings, labor savings, capital savings, or material savings – are often significant sources of economic benefit. The economic and energy benefits quantified herein are fuel or energy savings.

Environmental health benefits (co-benefits) are due to changes in the physical units of pollutants, focused primarily on changes in air emissions. Environmental health benefits may accrue through the reduction of adverse health events related to reductions in pollutant emissions associated with changes in the physical units of fossil-fuel energy consumed. The environmental health benefits quantified herein are health benefits associated with reduced emissions from driving EDVs relative to driving ICVs.

Energy security benefits refer to the reduced risks to the national energy infrastructure, increased national energy independence, and decreased exposure to exogenous (non-U.S.) volatility in fossil fuel trade.

Economic and energy benefits as well as environmental health benefits associated with VTO's R&D investments in NiMH and Li-ion battery technologies are quantified in monetary terms. Energy security benefits are described in quantitative and qualitative ways, but not in monetary terms.

Estimating economic and energy benefits

Economic and energy benefits are measured in terms of the fuel or energy savings associated with the diffusion of EDVs throughout the United States that are attributable to VTO's R&D investments. Economic and energy benefits are determined by the retrospective fuel savings associated with the actual diffusion of EDVs compared with the counterfactual diffusion of EDVs in the absence of VTO's R&D investments.

The monetized value of these counterfactually determined fuel savings are used as proxy for the economic benefits to society that are associated with VTO's R&D investments.[19] Part of these fuel-savings benefits are captured by individuals in the form of consumer surplus, and part of the economic fuel-saving benefits are captured by firms in terms of producer surplus (i.e., the generally higher prices that consumers pay for EDVs compared with traditional vehicles).

As discussed in detail in Chapter 4, informed individuals who were familiar with the supply chain for advanced battery development were asked to quantify changes in battery technology – battery life (years and charging cycles), energy density (Wh/kg), and cost ($/kWh) – that are attributable to VTO's R&D investments. Interviewees were then asked how those technology improvements could have affected the commercial viability and hence the diffusion of EDVs in the U.S. market. Specifically, interviewees were asked how a stylized adoption curve (showing the percentage of EDVs in the U.S. market increasing from 1999 through 2012) might have been different in the absence of VTO investments. This counterfactual question was posed by showing interviewees and survey respondents the adoption curve as a graphic object in a Word document that they could drag and/or reshape as they would expect it to have looked without VTO investments, with the actual adoption curve remaining fixed for reference. The relative difference between the actual and counterfactual adoption curves was applied to actual vehicle sales to estimate EDV sales attributable to VTO investments. Using extant information on average miles driven and average fuel economies of EDVs and conventional vehicles, attributable EDV sales were converted to attributable fuel savings.

Two important aspects of this approach must be addressed, the first being the identification of the next-best technology and the second relating to the attribution of benefits to VTO's investments in NiMH and Li-ion battery technologies.

One often defines the counterfactual situation in evaluation studies in terms of the next best alternative. That is, in the absence of public funding of a new technology under study, how would the existing technology have developed on its own? Following that line of reasoning, the next best alternative to an EDV with either a NiMH or Li-ion battery would have been a conventional ICV with a lead acid battery. VTO's R&D investments accelerated and enhanced the development and vehicle-specific application of NiMH and Li-ion battery technologies, thus accelerating the adoption of the technologies as commercialized innovations embodied in EDVs.

The attribution of social benefits is frequently a primary source of uncertainty in an evaluation study. Issues related to determining attribution often stem from obtaining multiple lines of evidence and from the extent to which that evidence comes from unbiased, independent sources. Data collection also presents challenges, such as lost or nonexistent records, key individuals who cannot be found or who choose not to respond to inquiries, and industry concerns about sharing proprietary information. Because the evaluation summarized in this book focuses on estimating the return on VTO's R&D investments in NiMH and Li-ion technologies, it is important to identify VTO's specific role in supporting technologies that led directly to society realizing over time the benefits described above.[20]

The identification of VTO's specific role in supporting the adopting of EDVs with NiMH and Li-ion batteries, as well as the impact of those technologies on market activities, was determined through detailed interviews with informed industry experts, as discussed at length in Chapter 4. In brief, attribution was definitional to the data collection approach. All information collection methods were carefully carried out in a manner such that the explicit data collected and the implicit insight gained were directly and specifically linked to quantifying and measuring social benefits with and without VTO's R&D investments in NiMH and Li-ion battery technologies.

Estimating environmental health benefits

Environmental health benefits attributable to VTO's R&D investments in NiMH and Li-ion battery technologies are quantified on the basis of attributable fuel savings. Attributable fuel savings are used as an input into the Co-Benefits Risk Assessment (COBRA) model, developed by the U.S. Environmental Protection Agency (EPA). The COBRA model provides estimates of health effects and their economic values that result from changes in the physical units of emitted pollutants. The COBRA model is discussed at greater length in Chapter 5.

Energy security benefits

Security impacts are measured in terms of the reduction of our nation's dependency on imported crude oil. Fuel savings from increased fuel economy are converted to gallons of crude oil saved and thus to the cumulative reduction of crude oil imported by the United States over the time period of the analysis.

Quantifying the social return

Economic evaluation metrics summarize the findings from an objective review, assessment, and comparison of program performance in a manner similar to any other financial investment analysis. In an economic evaluation in which all benefit streams are not quantified, and those that are quantified are truncated in time, it is important to emphasize that any performance measure is likely to be conservative and thus will understate the true net benefits to society.

Three measures of net social benefits are considered: net present value (NPV), benefit-to-cost ratio (BCR), and internal rate of return (IRR).

NPV, according to Circular A-94 of the Office of Management and Budget (OMB 1992), sets the standard evaluation criterion for deciding whether a government program can be justified on economic principles as the discounted monetized value of expected net benefits (i.e., benefits minus costs).[21] NPV is computed by assigning monetary values to benefits and costs, discounting future benefits and costs using an appropriate discount rate, and subtracting the sum total of discounted costs from the sum total of discounted benefits. Discounting benefits and costs transforms gains and losses occurring in different time periods to a common unit of measurement. From a social perspective, projects with positive NPV should generally be undertaken and those with negative NPV should generally not. Among those projects with positive NPVs, the larger the value of NPV the greater the net benefits to society.

BCR is the ratio of the present value of benefits to the present value of costs. Essentially, a BCR greater than 1 indicates that the present value of quantified benefits outweighs the present value of calculated costs. The larger the numerical value of a BCR, the greater the net benefits to society.

IRR is the discount rate that sets NPV equal to zero, or it is the discount rate that would result in a BCR equal to 1. The IRR's value can be compared with conventional rates of return for comparable or alternative investments. An IRR value greater than the return on an alternative investment (generally measured as equal to the discount rate) is interpreted to mean that the project was, in a comparative sense, socially valuable.

The specific formulae for these three measures are presented and discussed in Chapter 6.

Fundamental to the calculation of NPV and a BCR is the discount rate used to reference all values to the initial time period in which investment costs began. Following OMB (1992) guidelines, a 7 percent real (i.e., adjusted for inflation) rate of discount is used. The use of a real discount rate means that all measured benefits and all investment costs are converted into real, constant dollars to account for inflation. According to OMB (1992, p. 8):

> Constant-dollar benefit-cost analyses of proposed investments and regulations should report net present value and other outcomes determined using a real discount rate of 7 percent.

For comparative purposes, and following the more recent suggestion in OMB Circular A-4 (OMB, 2003), calculations of NPV and BCR using a 3 percent real rate of discount are also made and are presented in Chapter 6.[22]

Organization of the book

The remaining chapters in this book relate to an economic evaluation of the net social benefits attributable to VTO's R&D investments in NiMH and Li-ion battery technologies.

Chapter 2 places VTO's investment activities in the context of a public/private partnership model. The economic rationale for a public/private partnership is discussed along with frameworks for such partnerships that have been set forth in the academic and policy literatures. Institutional details about VTO's involvement, and that of private-sector companies, is summarized within the context of these frameworks.

Chapter 3 illustrates the adoption over time of EDVs that are powered by NiMH and Li-ion batteries. The adoption of EDVs as a proxy for the diffusion of NiMH and Li-ion batteries is a fundamental driver of the evaluation process.

Chapter 4 discusses in detail the measurement of economic and energy benefits associated with the adoption of EDVs over time. The counterfactual approach of quantifying the adoption of EDVs with and without VTO funding is described in detail, as is the methodology used to obtain through surveys data on the counterfactual state of being.

Chapter 5 discusses in detail the quantitative measurement of environmental health benefits and the qualitative consideration of energy security benefits. Environmental health benefits were quantified using the COBRA model; a discussion of that model is appended to the chapter.

Chapter 6 combines the streams of benefits presented in Chapters 4 (economic) and 5 (environmental/health) and compares them with the stream of VTO investments presented in Chapters 1 and 2, taking into account the time value of money to generate the conventional impact metrics: net present value, benefit-to-cost ratio, and internal rate of return. These metrics are calculated both for the full stream of benefits through 2022, taking into account the remaining useful lives of EDVs on the road as of December 31, 2012, and for the strictly retrospective stream of benefits, truncated on December 31, 2012. No part of the evaluation is based on any projection of future sales of EDVs.

Chapter 7 concludes the book with summary remarks and further discussion of some important assumptions, emphasizing that the economic impact estimates are conservative, lower-bound estimates of the social gains attributable to the VTO's R&D investments in NiMH and Li-ion battery technologies. Also presented in Chapter 7 are summaries of previous evaluation studies by EERE. These studies are noted in an effort to establish a benchmark for a relative comparison to EERE's investments in battery technology.

Notes

1 The Atomic Energy Commission was created by the Atomic Energy Act of 1946, Public Law 585 in the 79th Congress, to maintain control over atomic research and development. The Atomic Energy Act of 1954, Public Law 83-703, declared that "[a] tomic energy is capable of application for peaceful as well as military purposes," and thus the Atomic Energy Commission was given authority to regulate a commercial nuclear power. This separation of focus between government and commercial use of the atom was the precursor to the Energy Reorganization Act of 1974.

2 The OPEC oil embargo was not the first U.S. energy shortage. Some shortages were realized in the "great blackout" of 1965 – a disruption of electric service in Ontario, Canada and Connecticut, Massachusetts, New Hampshire, Rhode Island, Vermont, New York, and New Jersey in the United States on November 9, 1965, due to human error – and several brownouts in 1971. For details, see Fehner and Holl (1994) and the George Mason University Blackout History Project (http://blackout.gmu.edu/). President Nixon "warned that the United States could no longer take its energy supply for granted. Since 1967, Nixon observed, America's rate of energy consumption had outpaced the Nation's production of goods and services [and] he asked Congress to establish a department of natural resources to unify all important energy resource development programs" (Fehner and Holl, 1994, pp. 4–5).

3 For more information, see: http://uscode.house.gov/download/pls/15C52.txt.

4 This terminology is discussed again in Chapter 3.

5 EPAct reaffirmed this mandate and authorized the Secretary of Energy to join cooperative agreements with industry to develop advanced batteries for EVs (NRC 1998).

6 According to NRC (1998), EPRI joined the consortium because of its long history of research in batteries and because the use of batteries in EVs would allow utilities to use their excess capacity during off-peak hours.

7 The scope of the Partnership for a New Generation of Vehicles (PNGV) broadened over time. For example, pursuant to the National Cooperative Research and Production Act of 1993, Public Law 103-42, which amended the National Cooperative Research Act of 1984 (Public Law 98-462), the Environmental Research Institute of Michigan gave written notice on August 5, 1996 to the Antitrust Division of the U.S. Department of Justice and to the Attorney General and the Federal Trade Commission, that Case Western Reserve University, Cleveland, OH; Chrysler Corporation, Auburn Hills, MI; Delaware Machinery and Tool Company, Inc., Muncie, IN; Doehler Jarvis, Toledo, OH; EDCO Engineering, Toledo, OH; Environmental Research Institute of Michigan, Ann Arbor, MI; Ford Motor Company, Dearborn, MI; General Motors Corporation, Warren, MI; Ohio State University, Columbus, OH; and Prince Machine, Holland, MI would conduct joint research under the direction of the PNGV on improvements in the efficiency of aluminum die casting operations to determine the causes of porosity in transmission cases and modifications defined for the production process.

8 The PNGV's progress in battery technology is reviewed in NRC (2001). See also Trinkle (2009).

9 See http://energy.gov/eere/about-us. Other program offices within DOE are: Advanced Research Projects Agency for Energy, Loan Program Office, Office of Electricity Delivery and Energy Reliability, Office of Environmental Management, Office of Fossil Energy, Office of Indian Energy Policy and Programs, Office of Legacy Management, Office of Nuclear Energy, and the Office of Science. See, http://energy.gov/offices.

10 See, http://energy.gov/eere/about-us/mission.

11 See, http://www1.eere.energy.gov/site_administration/programs_offices.html.

12 See, http://www1.eere.energy.gov/vehiclesandfuels/about/fcvt_mission.html.

13 See, http://www1.eere.energy.gov/vehiclesandfuels/technologies/energy_storage/index.html.

14 See www.nrel.gov/vehiclesandfuels/energystorage/resources.html?print. Although the focus of the evaluation discussed in this book is on light-duty vehicles, much of VTO-sponsored R&D will transfer to energy storage for heavy-duty hybrid vehicles as well.

15 Link and Scott (2010, pp. 28–31) provide a critical discussion of the Griliches/Mansfield approach. Drawing on Link and Scott (1998, 2010), alternative methodologies for an evaluation might be considered. Whereas the Griliches/Mansfield methodology takes the counterfactual to be the status-quo technology that prevailed prior to the new technology brought forth by public R&D investments, our approach takes into account the realistic possibility that the private sector would have made some progress even absent the public investment. Link and Scott (1998) carefully treat the distinctions between this approach and the most traditional Griliches/Mansfield approach, including two important special cases: If the private sector would not have attempted to replace at all the progress made with the support of the public sector, we have the traditional Griliches/Mansfield approach. If instead the private sector would have completely replaced the technological progress, the question to ask is: What would it have cost the private sector to do so? In this case, the public investment is justified if it is a more efficient – i.e., lower-cost – way of achieving the same outcome. Realistic scenarios

will typically fall between these two special cases: in the counterfactual without public investment, it is natural to suppose that both the stream of private investments in related technologies and the technical development of those technologies (and associated streams of social surplus created as those technologies are adopted) would have been different. The ideal evaluation would accurately quantify the actual and counterfactual streams of social surplus, net of the streams of private R&D expenditures in each case, and compare the difference in the present value of these streams with the present value of the public R&D expenditure. This ideal approach, in which nothing is held constant, is difficult to implement in practice. It is often more practical to begin with one of the special cases (holding constant either the technical advancements or the private investment), then determine which assumptions are most important to relax, and do so (thus approaching more closely the ideal) to the extent feasible. For the present study, there was widespread agreement among interviewees that battery technology would have made substantially slower progress without VTO support. There was less clear consensus on whether the private sector would have invested more or less had VTO investments not been made. Therefore, of the two special cases, the Griliches/Mansfield approach is closest to the most realistic counterfactual. However, it also seemed unlikely to most (although not all) interviewees that the status quo as of 1992 – that of no EDVs on the U.S. auto market – would have prevailed without VTO investments. Therefore we adopted a counterfactual in which EDVs could enter the U.S. market (as attributes of battery technology developed) at a rate different than that actually observed. Either of the special cases – of the private sector without VTO investments either completely or not at all replicating the technical progress that was actually made with VTO investments – could of course be offered by interviewees, but such responses were rare. In the absence of strong views among our respondents on whether private investment would have been greater or less in the absence of VTO investment, it was deemed most appropriate to hold private investment constant.

Link and Scott (2010) discuss another alternative methodology: spillover analysis, which attempts to quantify the positive externality associated with spillovers of knowledge generated by R&D investment. Part of the appeal of this approach is that it directly addresses a principal justification for public investment: that the private sector would not have made the socially productive investments because too little of the social value generated could be appropriated by the investing company or consortium of companies. Quantifying the value of knowledge spillovers from investments in battery technology was not deemed practical for this study. That such spillovers undoubtedly exist but were not quantified and included in the stream of benefits implies an element of caution in this evaluation.

16 The approach adopted for this study follows EERE guidelines for retrospective benefit-to-cost studies (Ruegg and Jordan 2011). EERE has adopted these guidelines to promote consistency and comparability of evaluations of its programs. The motivation for evaluations, like this one, of public investments is rooted in legislation, specifically the Government Performance and Results Act (GPRA) of 1993, Public Law 103-62. As Link and Scott (2010) discuss in detail, GPRA builds upon the February 1985 GAO report and the Chief Financial Officers Act of 1990. The 103rd Congress stated in the August 3, 1993 legislation that it found, based on over a year of committee study, that:

1 waste and inefficiency in Federal programs undermine the confidence of the American people in the Government and reduce the Federal Government's ability to address adequately vital public needs;

2 federal managers are seriously disadvantaged in their efforts to improve program efficiency and effectiveness, because of insufficient articulation of program goals and inadequate information on program performance; and
3 congressional policymaking, spending decisions and program oversight are seriously handicapped by insufficient attention to program performance and results.

Accordingly, the stated purposes of GPRA are to:

1 improve the confidence of the American people in the capability of the Federal Government, by systematically holding Federal agencies accountable for achieving program results;
2 initiate program performance reform with a series of pilot projects in setting program goals, measuring program performance against those goals, and reporting publicly on their progress;
3 improve Federal program effectiveness and public accountability by promoting a new focus on results, service quality, and customer satisfaction;
4 help Federal managers improve service delivery, by requiring that they plan for meeting program objectives and by providing them with information about program results and service quality;
5 improve Congressional decision making by providing more objective information on achieving statutory objectives, and on the relative effectiveness and efficiency of Federal programs and spending; and
6 improve internal management of the Federal Government.

17 The Program Office at the National Institute of Standards and Technology (NIST) within the U.S. Department of Commerce might arguably be credited with pioneering the art of program evaluation of publicly funded and privately performed R&D along the lines used in our study of battery technologies (Link and Scott, 2012). The various complementary approaches discussed in Link and Scott (2012) may be described as attempts to implement, within the practical constraints of program evaluation, the conceptual ideal described by Tassey (2003, p. 15), then Director of the Program Office at NIST:

> An ideal analytical approach [for a retrospective evaluation] is the construction of a time series of economic activity of affected industries that includes a period before government intervention. At some point in the time series a government funded project . . . occurs and the subsequent portion of the time series reflects the technical and economic impacts of the intervention.

18 Hufschmidt (2000) offers an interesting historical perspective about benefit–cost analysis – and we view the evaluation metrics used in this book as falling broadly under the conceptual umbrella of benefit–cost analysis. According to Hufschmidt (2000, p. 42): "The issue of applying economic benefit–cost test to public investment projects first arouse in the United States (U.S.) during the great depression of the 1930s. . . . [With President Franklin Roosevelt's massive programs of public works] the question soon arose on how to assess the social worth or value of individual projects." Hufschmidt claims that the issue of how to assess social worth of a project was addressed by the National Planning Board in 1934. The Board commissioned a study

by Professor Clark at Columbia University. His approach was based on a willingness to pay concept. The U.S. Bureau of the Budget released Circular A-47 in 1952. The circular contained basic guidelines about what to include as benefits and what to include as costs. Interesting is that the Circular called for the use of a discount rate that was to long-term government bonds, and prospective analyses were focused on a 50-year time horizon (see Note 21 below).

19 We realize that there are other forms of social surplus generated by the VTO's investments, such as knowledge spillovers to other industries

20 This is often referred to as program additionality.

21 Quoting from OMB Circular A-94 (1992, p. 3): "The standard criterion for deciding whether a government program can be justified on economic principles is net present value – the discounted monetized value of expected net benefits (i.e., benefits minus costs). Net present value is computed by assigning monetary values to benefits and costs, discounting future benefits and costs using an appropriate discount rate, and subtracting the sum total of discounted costs from the sum total of discounted benefits. Discounting benefits and costs transforms gains and losses occurring in different time periods to a common unit of measurement. Programs with positive net present value increase social resources and are generally preferred. Programs with negative net present value should generally be avoided."

22 For federal economic evaluations, OMB issues directives on discounting and discount rates for different types of evaluations. Circular A-94 (OMB 1992) directs the use of a 7 percent real discount rate for federal benefit–cost analysis. More recent guidance is provided by Circular A-4 (OMB 2003), which pertains to benefit–cost analysis used as a tool for regulatory analysis. It notes that Circular A-94 stated that a real discount rate of 7 percent should be used in benefit–cost analysis as an estimate of the average before-tax rate of return to private capital in the U.S. economy. This rate is an approximation of the opportunity cost of capital. Circular A-4 further notes that OMB found in a subsequent analysis that the average rate of return to capital remained near 7 percent. It also points out that Circular A-94 recommends using other discount rates to show the sensitivity of the estimates to the discount rate assumption and notes that the average real rate of return on long-term government debt has averaged about 3 percent. Circular A-4 requires the use of both a 7 percent and a 3 percent real discount rate for a benefit–cost analysis conducted for regulatory purposes. When regulation primarily and directly affects private consumption (e.g., through higher consumer prices for goods and services), a lower discount rate is appropriate, and OMB suggests a 3 percent real rate of time preference. For the purpose of discounting constant dollar cash flows in this study, both rates are used – a 7 percent and a 3 percent real discount rate – even though the purpose is not regulatory.

2 Public/private research partnerships

This chapter begins with a discussion of the economic concept of market failure as the theoretical mandate for public sector intervention into private sector activities. The concept of market failure is the justification offered for public sector support of science and technology, or R&D, such as, in the present study, for battery technologies for electric drive vehicles. Then, having offered an economic argument for DOE's investments in battery storage technologies to be utilized in the private sector, the relationship between DOE and the private sector is described in terms of a public/private partnership.

Market failure and private investments in R&D

Market failure describes a situation where market forces lead to an inefficient allocation of resources from a social perspective. Public investments in science and technology (or R&D) are justified in principle by their potential to correct market failures involving underinvestment by the private sector in basic and applied research and the development and commercialization of new technology. Martin and Scott (2000, p. 438) observe about market failure:

> Limited appropriability, financial market failure, external benefits to the production of knowledge, and other factors suggest that strict reliance on a market system will result in underinvestment in innovation, relative to the socially desirable level. This creates a *prima facie* case in favor of public intervention to promote innovative activity.

These and other factors contributing to private-sector underinvestment in R&D (termed barriers to technology and innovation) have been elaborated on by Link and Scott (2010); thus, only a brief overview follows.[1]

First, the social rate of return to R&D investment is likely to exceed the private rate of return for several reasons. The scope of potential markets for new technology is often broader than the market strategy of any one firm, making it unlikely that one firm could appropriate (even if it could envision) all of the social returns to its R&D, particularly for activities that tend toward the more basic end of the research spectrum. Knowledge and ideas generated by one firm's R&D investments will often spill over to other firms, both rivals competing in the same markets and (because of the breadth of application for new technology) firms in unrelated markets. Such spillovers are socially valuable but not privately rewarding to the firm making the R&D investment. A firm may also anticipate some amount of opportunism by potential buyers of the new technology it develops; a potential buyer may learn enough about the new technology, in the process of making its decision whether to buy, that it can invent around any intellectual property protections and acquire the value of the new technology without paying for it.[2]

Second, the private hurdle rate (the expected rate of private return that an R&D investment must meet to be deemed worthwhile to the firm) is likely to be higher than the social hurdle rate (the opportunity costs of the public's investment funds). Given the technical and commercial risks associated with R&D investments (will the R&D achieve certain technical criteria and will the market embrace the resulting innovation?), owners and lenders who provide the investment capital for R&D will generally require a higher risk premium than will society, if for no other reason than that society can spread its R&D investments over a larger portfolio. Even small differences between the private and social costs of investment capital (in terms of a required annual expected rate of return) can lead to substantial differences between the private and social net present values of R&D projects when there is a long lag between the time that investments are made and the time that returns are realized. For example, assuming a 10 percent private and 5 percent social cost of capital, the ratio of private to social present value of a future return is 0.79 five years into the future, 0.63 ten years out, and 0.39 twenty years out.[3]

Figure 2.1 illustrates the implications of these two observations – that social returns to R&D investments typically exceed private returns and that private hurdle rates typically exceed social hurdle rates – for public policy decisions involving R&D investment.[4] R&D projects for which the social rate of return exceeds the private rate of return lie above the 45-degree line in Regions I, II, and III. Projects in Region I are neither privately nor socially valuable: their rate of return is less than the hurdle rate from both the private and social perspective. Projects in Region II (like Project A) are good candidates for public investment: the social rate of return for these projects exceeds the social hurdle rate, and their private rate of return falls short of the private hurdle rate – meaning that these projects are socially valuable but unlikely to be undertaken by the private sector. Projects in Region III (like Project B) are socially valuable but poor candidates for public investment: since the private rate of return to these projects exceeds the private hurdle rate, we would expect them to be undertaken by the private sector; public investment in such projects would only crowd out the private investment.

Conflated in the private and social rates of return are the returns themselves – the streams of private and social gains ensuing from an investment in R&D – and the costs – the streams of R&D expenditure. The reasons given above for why the social rate of return may exceed the private rate of return focused on the arguably more straightforward reasons why the streams of social returns may be greater than the streams of private returns. In fact there are also compelling arguments for why public investments in R&D, especially in collaboration with

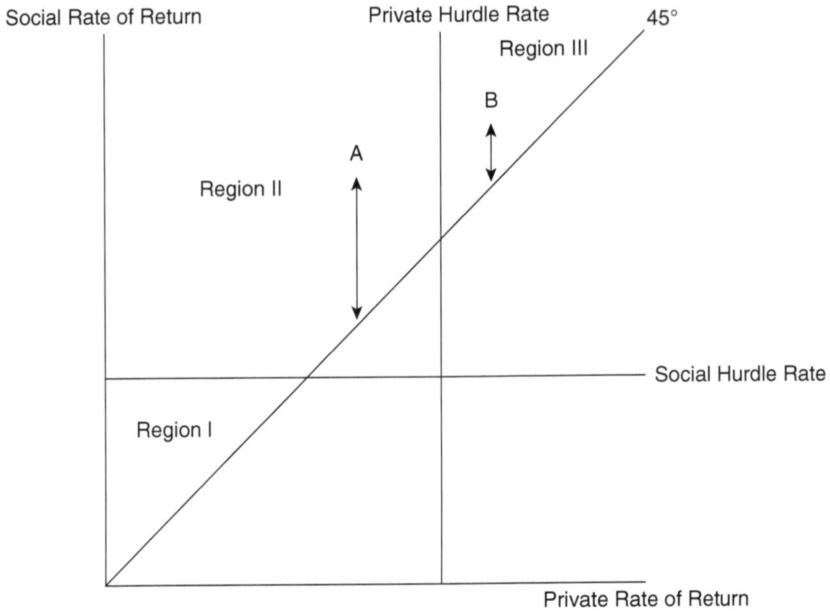

Figure 2.1 Decision-making model for public R&D investments

private-sector partners, may lower the expected cost of achieving a given R&D outcome. A successful R&D effort may, for example, require (Link and Scott, 2010, p. 9):

> multidisciplinary and multiskilled research teams; unique research facilities not generally available within individual firms; or fusing technologies (i.e., technologies used together or sequentially) from heretofore separate, non-interacting parties [or] investments in combinations of technologies that, if they existed, would reside in different industries that are not integrated.

In such cases (Link and Scott, 2010, p. 10):

> [u]nderinvestment will occur not only because of the lack of recognition of possible benefit areas or the perceived inability to appropriate whatever results but also because coordinating multiple players in a timely and efficient manner is cumbersome and costly. . . . [S]ociety may be able to use a technology-based public institution to act as an honest broker and reduce costs below those that the market would face.

The argument for public investment in R&D to bring forth new technologies that improve environmental outcomes is especially strong because of the negative externalities associated with pollution (Jaffe et al., 2005). In the case of battery technology for electric drive vehicles, drivers underappreciate the social value of the more fuel-efficient technology because the price they pay at the pump does not fully reflect the social opportunity cost of gasoline; namely, the price of gasoline does not reflect the cost of the associated pollution. Therefore, the gap between the private and social returns to R&D investments in technologies that improve vehicle fuel efficiency, including the battery technologies considered in this book, is likely to be especially large.

A public/private partnership model

Consider the following definition of aspects of a public/private partnership model. Link (1999) first proposed this definition and then Link and Link (2009, p. 32) elaborated on it:

> The term *public* refers to any aspect of the innovation [and technology development] process that involves the use of governmental [i.e., public-sector] resources, be they federal, state, or local in origin. *Private* refers to any aspect of the innovation [or technology development] process that involves the use of private-sector resources, mostly firm-specific resources. And, *resources* are broadly defined to include all resources – financial resources, infrastructural resources, research resources, and the like – that affect the general environment in which innovation [or technology development] occurs. Finally, the term *partnership* refers to any and all innovation-related [and technology-development-related] relationships, including but not limited to formal and informal collaborations or partnerships in R&D.

The framework that defines our view of a public/private partnership is described in Table 2.1. The first column of the table describes the nature and scope of the public sector's involvement in the partnership. The public sector's involvement can be indirect or direct, and if direct there could be an explicit allocation of public resources including financial, infrastructural, and/or research resources. The second and third columns in the table relate to the

Table 2.1 Taxonomy of public/private partnership mechanisms and structures

	Economic objective	
Public sector involvement	*Leverage public-sector R&D*	*Leverage private-sector R&D*
Indirect	. . .	. . .
Direct		
Financial resources	. . .	. . .
Infrastructural resources	. . .	. . .
Research resources	. . .	. . .

economic objectives of the public/private partnership. Broadly, the objectives are either to leverage public-sector R&D activity or to leverage private-sector R&D activity. Although the objectives of innovative activity are often multifactorial, for illustrative purposes, one single overriding economic objective is assumed here.

Public resources involved in battery technology R&D, or more precisely energy storage R&D, came from the VTO. These resources are direct. VTO's financial contributions are shown in Figure 1.1 to have begun in 1976. But also, significant research on battery technology occurs at DOE's national laboratories so there is an infrastructural resource and research resource contribution as well.

Regarding private resources, an important part of VTO's support for NiMH and Li-ion battery technologies was the funding of U.S. Advanced Battery Consortium (USABC). USABC shared the cost of contracts to private-sector companies to conduct in-house research on battery technology.[5]

DOE's Office of Transportation Technologies (OTT) managed USABC's contracts to private companies; these contracts were awarded through a competitive process. The OTT also managed Cooperative Research and Development Agreements (CRADAs) with DOE's national laboratories.[6] These CRADAs often focused on developing test procedures and evaluating batteries developed in the USABC program.

A research director with a company that now manufacturers Li-ion batteries for EDVs described the company's present chemistry as being a direct result of interaction with DOE, through USABC, and emphasized the importance of this interaction in leveraging for vehicle applications the company's prior work on smaller platforms: cell phones and laptops, where (to paraphrase) "the protocols and formalisms used to test the batteries are entirely different." As an example, "the development of vehicle batteries requires you to think of a 10–15 year lifespan compared to 1–3 years for cell phones and laptops." To paraphrase another respondent, "uniform technical targets for vehicle-batteries were set by the USABC program and these formed the basis for several of the manufacturers and original equipment manufacturers (OEMs) to develop battery chemistries that are durable through the life of the vehicle." In the words of another, "without the VTP push for cycle and calendar life, there was very little incentive for the battery to reach 3000 cycles and a 10-year life; consumer electronics require only 500 cycles and a 2-year life, and can tolerate high price."

Other respondents emphasized the importance of USABC in developing human capital in the United States. One industry respondent said, "DOE funding has led not only to technical breakthroughs but has also led to an accumulation of human capital and expertise in the U.S., which would not exist at all without DOE funding; without the DOE, the advancements that have been made in U.S. industry and government and university labs over the past 20 years would have taken 30 years longer – what has taken 20 years would have taken 50 without DOE."[7] Said another respondent, "without VTP, a whole generation of battery scientists would have been lost, given the state of funding for battery research in the 1990s; although the funding level was still low, at least it was maintained, and when interest in batteries exploded in the mid-2000s, those students, who had by then taken jobs at universities and companies, were there to re-ignite the research in the U.S."

As shown in Figure 2.2, DOE invested $315 million in 2012 dollars in energy storage technologies through its funding of USABC contracts from 1992 through 2010 (2010 is the last year of available information on USABC contracts). Private-sector R&D investment amounted to an additional $358 million in 2012 dollars over the same period.

Approximately 9 percent of VTO's total USABC R&D investments supported NiMH battery research. U.S. companies receiving support for NiMH R&D included: Energy Conversion Devices, Inc. (ECD), also known as ECD Ovonic; Ovonic Battery Company, Inc., a subsidiary of ECD Ovonic; GM Ovonic, a joint venture between GM and Ovonic Battery Company; Texaco Ovonic Battery Systems, a joint venture between Texaco and Ovonic Battery Company (Texaco acquired GM's interest); and Cobasys LLC, a joint venture between Chevron and Ovonic Battery Company (Chevron acquired Texaco's interest). To paraphrase one industry respondent, "NiMH technology was around in the 1980s, but not until the 1990s did Ovonic start scaling up for large capacity suitable for vehicle applications; Ovonic was a small company and could not have done what it did without DOE and USABC."

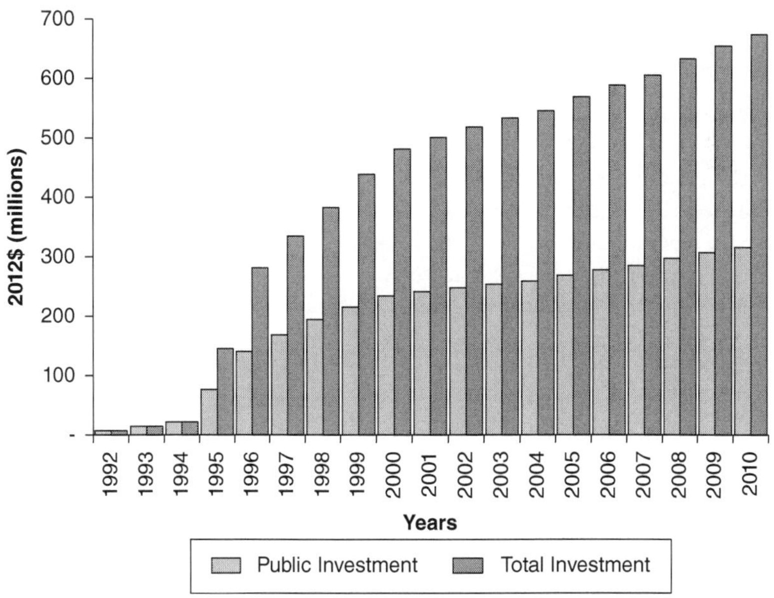

Figure 2.2 Cumulative USABC R&D investments in energy storage technologies, 1992 through 2010

Approximately 50 percent of VTO's total USABC investments supported Li-ion battery research. U.S. companies receiving support for NiMH battery research included: 3M; Johnson Controls; Saft; Johnson Controls/Saft (JCS), a joint venture between Johnson Controls and Saft; A123Systems; and LG Chem Power Inc., a North American subsidiary of LG Chem Ltd. See Figure 2.3 for a visual description of the timeline for VTO's support of these companies. To paraphrase one respondent, whose work in industry had leveraged research performed in government laboratories,

> the impact of VTO funding has been significant in that it allowed the emergence of new types of materials that would enhance the energy density of Li-ion batteries; without VTO funding, the work that Argonne National Labs and Lawrence Berkeley Labs have done on new materials would not have been possible; VTP funding has firmly established the battery industry in the U.S. right now, so that battery manufacturing can happen in the U.S. – plants are being built, machinery has been installed, and the U.S. can be a credible supplier of batteries for vehicle applications.

Another industry respondent said of VTO's impact, "without DOE, there would be essentially no U.S. industry; technology would still have been developed abroad in Japan and Korea, for example, and EDVs would still have made their way into the U.S. market, but it would have taken longer – perhaps 10 years longer – to get to where we are today, and there would presently be no Li-ion batteries on the road."

In addition to funding the private sector for research on battery technologies, USABC's support of private-sector research established what have become the standardized performance metrics for batteries. See Table 2.2 for definitions of these performance metrics.

Thus, the public/private partnership framework in Table 2.1 can be filled in to reflect the resource contribution of the public sector (i.e., VTO) and the private sector (i.e., recipient companies through USABC); see Table 2.3.

Table 2.3 shows that DOE's role in the public/private partnership to support the development of new battery technology, NiMH and Li-ion battery technologies in particular, was direct. DOE, and VTO in particular, contributed financial, infrastructural, and research resources to leverage private-sector R&D. In addition, DOE provided direct financial support to leverage ongoing battery technology research at the national laboratories.

This framework for the mechanisms and structure of a public/private partnership complements an earlier conceptualization by the Office of Technology Policy (OTP, 1996). The OTP's taxonomy classified public/private partnerships in the United States at two undefined points of transition to emphasize the evolution of the government's role – from that of a customer of private-sector research to that of a partner in that research. Specifically:

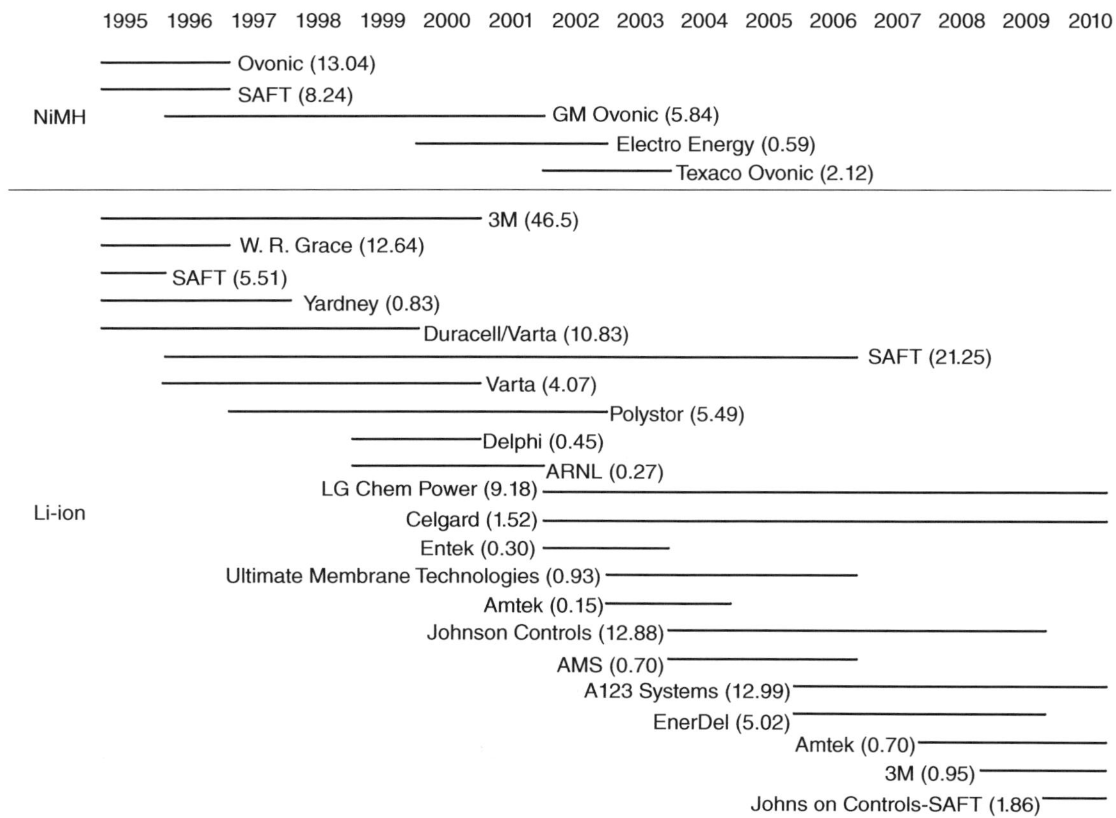

Figure 2.3 VTO's R&D investments for NiMH and Li-ion battery technologies, by company, 1995 through 2010 (millions $)

Table 2.2 Performance metrics established by USABC

Metric	*Definition*
Specific energy (Wh/kg)	A measure of the total energy density of the battery pack per unit weight. It provides an indication of the vehicle range. Total energy is analogous to the size of the gas tank on a conventional, combustion-engine-powered automobile. This metric is important for EV batteries because added mass requires more energy to move.
Energy density (Wh/L)	A measure of the total energy stored in the battery pack per unit volume. This metric is important in portable electronics where size is often limited.
Specific power (W/kg)	A measure of the total power that can be delivered per unit weight. Power, which is energy divided by the time it is delivered, translates to the acceleration ability of the source. In EVs, power is limited by how fast the energy in a battery pack can be delivered to motors or electrical circuitry.
Power density (W/L)	A measure of the total power per unit volume of the battery pack that can be delivered in a short burst of time such as during acceleration.
Life (years)	An engineering estimate of the expected time that an EV battery pack can be fully charged and discharged and maintain a specified capacity threshold. Some degradation in the pack's specific energy occurs over time, and the battery needs to be replaced when its capacity falls below a specified percentage of the original value.
Cycle life (cycles)	The number of times a battery pack can be charged and discharged. Each charge-discharge event constitutes one cycle. Typically cycle life is related to the depth of discharge (DOD) of the battery pack. Deep discharge and charge cycles will lower cycle life.
Ultimate price ($/kWh)	The cost of a battery pack in dollars divided by the total energy that is contained (in kWh) in a single charge of the battery.
Operating environment	The environmental conditions under which the battery pack is expected to operate. The operating environment usually consists of a lower temperature limit and an upper temperature limit. Batteries typically do not operate in extremely cold conditions and may exhibit reduced performance or become unsafe at high temperatures.
Capacity	The total energy stored in the battery, usually expressed in kWh.
Recharge time (hours)	The time that it takes to recharge the battery to a predetermined acceptable level. Recharge time is often expressed as C rate. A recharge rate of 1C implies that the full capacity of the battery can be restored after 1 hour of charging, whereas a recharge rate of 0.1C implies that it takes roughly 10 hours to fully charge the battery pack.
Continuous discharge in 1 hour	Energy delivered in a constant power discharge required by an EV for hill climbing and high-speed cruising, specified as the percentage of energy capacity delivered in a 1-hour constant power discharge.
Power and capacity degradation	Performance degradation defines the extent to which the battery system is unable to meet the original performance specification.

Note: Wh = watt-hours, L = liter, kg = kilogram.

Table 2.3 DOE's role in the public/private partnership to support battery technology

Public sector involvement	Economic objective: Leverage Public-sector R&D	Leverage Private-sector R&D
Indirect	. . .	. . .
Direct		
Financial resources	yes	yes
Infrastructural resources	. . .	yes
Research resources	. . .	yes

> By the late 1980s, a new paradigm of technology policy had developed. In contrast to the enhanced spin-off programs – enhancements that made it easier for the private sector to commercialize the results of mission R&D – the government developed new public-private partnerships to develop and deploy advanced technologies. . . . [T]hese new programs . . . incorporate features that reflect increased influence from the private sector over project selection, management, and intellectual property ownership. Along with increased input, private sector partners also absorb a greater share of the costs, in some cases paying over half of the project cost. . . . The new paradigm has several advantages for both government and the private sector. By treating the private sector as a partner in federal programs, government agencies can better incorporate feedback and [thus] focus programs. Moreover, *the private sector as partner* [emphasis added] approach allows the government to measure whether the programs are ultimately meeting their goals: increasing research efficiencies and effectiveness and developing and deploying new technologies.
>
> (OTP, 1996, pp. 33–34)

Figure 2.4 illustrates this OTP view. There are several salient features in Figure 2.4. First, the federal government has changed from being a customer for

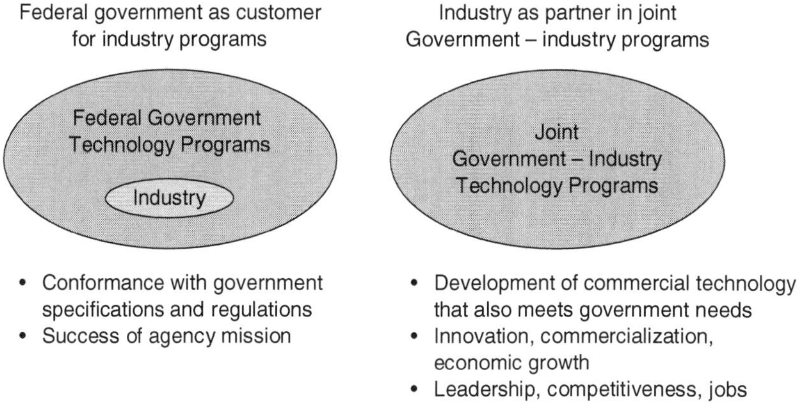

Figure 2.4 Innovative paradigm for a public/private technology partnership

Source: Based on Office of Technology Policy (1996, p. 34).

the technology output of industry programs, which it often financed, to a partner, and often a research partner, in the programs. And second, not only does this role change increase the ability of industry to focus its efforts more efficiently on government needs, but also it speeds up the technology diffusion process.

In the following chapter we introduce advancements in NiMH and Li-ion battery technologies that resulted from VTO's R&D support of those technologies. We argue that improvements in the performance characteristics of the state-of-the-art technology have translated over time into the commercial availability of EDVs and the subsequent market adoption of EDVs. Chapter 3 summarizes the market adoption of NiMH and Li-ion battery technologies through the purchase of EDVs in the United States beginning in 1999. This summary then becomes the benchmark for the evaluation analysis that follows in later chapters.

Notes

1 The discussion of specific barriers to innovation and technology by Link and Scott (2010, pp. 8–11) follows the foundational work of Kenneth Arrow (1962, p. 609), which pointed out that invention, or the production of knowledge, is characterized by "three of the classical reasons for the possible failure of perfect competition to achieve optimality in resource allocation: indivisibilities, inappropriability, and uncertainty."

2 Potential buyers, for their part, may worry that the seller of a new technology may learn enough about the buyer's operations that it could "back away from the transfer and instead enter the buyer's industry as a technologically sophisticated competitor" (Link and Scott, 2010, p. 11).

3 Taking the cost of capital (as a required annual rate of return) to be r, the present value of a return, x, realized t years into the future, is $x/(1+r)^t$. Thus the ratio of private to social present value in our example is $(1.05/1.1)^t$.

4 Figure 2.1 is adapted from Link and Scott (2010, p. 6) and Jaffe (1998, p. 16).

5 USABC investments in NiMH and Li-ion technology began in earnest in 1995. However, USABC was established in 1991 and served as a model for other public/private partnerships that came after. Most notable among these in the U.S. automotive industry was the Partnership for a New Generation of Vehicles (PNGV), launched in 1993 by the Clinton Administration and the "Big Three" U.S. automakers: Ford, Chrysler, and General Motors. See Sperling (2001) and Trinkle (2009). Trinkle notes that "not only did USABC and other consortia precede PNGV, but their preexistence facilitated its conception and implementation" (p. 55).

6 The Stevenson-Wydler Technology Innovation Act of 1980, Public Law 96-480, called for federal laboratories to actively promote technology transfer to the private sector for commercial exploitation. In 1986, the Stevenson-Wydler Act was amended by the Federal Technology Transfer Act of 1986, Public Law 99-502, to, among other things, enable the laboratories to enter into CRADAs with outside organizations or parties:

> Each Federal agency may permit the director of any of its Government-operated Federal laboratories – (1) to enter into cooperative research and development state and local agreements on behalf of such agency . . . with other Federal agencies; units of State or local government; industrial organizations (including corporations, schools and partnerships, and limited partnerships, and industrial development organizations); public and private foundations; nonprofit organizations

(including universities); or other persons (including licensees of inventions owned by the Federal agency); and (2) to negotiate licensing agreements . . . .

In subsequent legislation, the activities that fall under a CRADA and related licensing arrangements were expanded. These legislations included the National Competitiveness Technology Transfer Act of 1989, Public Law 101-189; the National Technology Transfer and Advancement Act of 1995, Public Law 104-113; and the Technology Transfer Commercialization Act of 2000, Public Law 106-404.

7 In this, as in all instances where quotation marks are applied to the comments of interview respondents, only paraphrasing is implied.

3 The adoption of battery technology in EDVs

Advancements in NiMH and Li-ion battery technologies – that is, improvements in the performance characteristics of the state-of-the-art technology – have translated over time into the commercial availability of EDVs and the subsequent market adoption of EDVs. VTO has influenced the diffusion of these technologies not only through its investments in the development of the technologies to make them more commercially viable but also through ancillary outreach activities. Said one respondent (to paraphrase), "VTO projects generated tremendous indirect effects by their outreach efforts through workshops, seminars and demonstrations to educate the general public about these new technologies, which have significantly increased the market acceptance of EDVs powered by NiMH and Li-ion batteries." Other respondents emphasized the link between the technical progress VTO supported and market adoption of EDVs that can therefore be attributed to VTO. Said one, "Shorter cycle life and lower energy density coupled with higher cost resulting from less R&D investment would make Li-ion less viable." Said another, "Without the VTO's support of Li-ion, I think EDVs would be more heavily NiMH, with the only Li-ion powered vehicles coming from Asia; VTO support of A123 and Compact Power led to the acceleration of commercialization of the Chevy Volt and other smaller producers, like Fisker and Tesla." Still another industry respondent thought it possible that without the VTO's support for battery technology development, "There might presently be no Li-ion technology in the EDV market." Said this respondent, "The influence on U.S. automakers is evident with all offering, or very soon to offer, EDVs exclusively with Li-ion technology; by contrast, the Japanese automakers have been relatively slow to adopt Li-ion technology in vehicle applications, and still are not offering it in their HEV models." This chapter summarizes the market adoption of NiMH and Li-ion battery technologies through the purchase of EDVs in the United States beginning in 1999.[1]

In our opinion, a technical understanding of battery technology is not requisite for an understanding of the evaluation of VTO's R&D investments in battery technology or for an appreciation of the implications from our evaluation. For completeness, however, we provide a technical overview of battery technology, with particular emphasis on NiMH and Li-ion batteries, in the appendix to this chapter.

To repeat from Chapter 1, we use the term *EDVs* to describe all types of EDVs, including:

- hybrid electric vehicles (HEVs), which use gasoline to charge the battery and part of the time to power the vehicle (e.g., the first-generation Prius);
- electric vehicles (EVs), which are powered exclusively by a battery and must be plugged into an electrical outlet to recharge (e.g., the Tesla Model S, Nissan Leaf); and
- plug-in hybrid electric vehicles (PHEV), which can either use gasoline to recharge the battery and power the vehicle or be plugged in to recharge the battery (e.g., the Chevrolet Volt).

In the United States, 113,257 EDVs (all HEVs) were sold from 1999 through 2003. By the end of 2007, over 1 million EDVs had been sold. This number includes the sales from 2004 through 2007 of 200,231 hybrid sport-utility vehicles and light-duty trucks and 450,080 Toyota's Prius. All of these vehicles were hybrids using NiMH batteries.

Annual sales of EDVs began to increase in 2000 and then peaked at 351,071 in 2007; sales faltered during the Great Recession (December 2007 to June 2009). In contrast, EDVs as a percentage of all U.S. car and truck sales peaked in 2009. The recent rebound has been striking, with EDV sales reaching 487,890 in 2012 and 592,287 in 2013, as shown in Figure 3.1. This more recent trend was helped

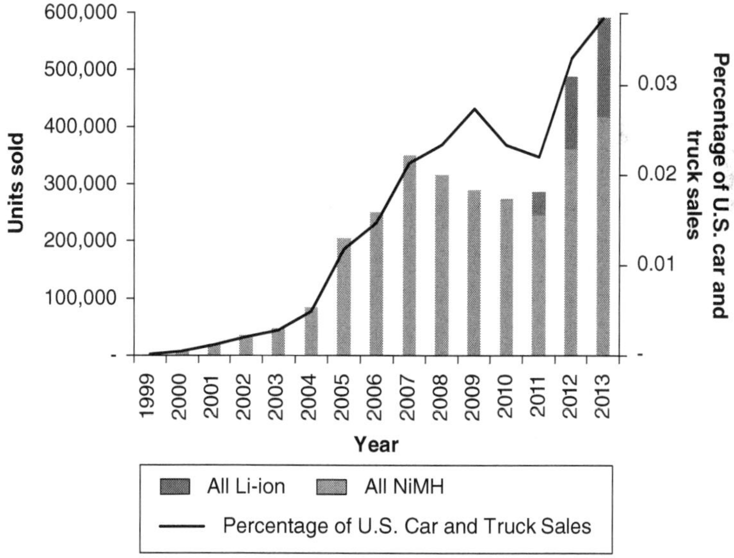

Figure 3.1 Electric drive vehicles in the United States, by battery technology and by type, 1999 through 2012

Source: This figure combines the EDV sales data from Tables 3.1 through 3.3 with U.S. car data from Wards Auto (http://wardsauto.com/keydata/historical/UsaSa01summary).

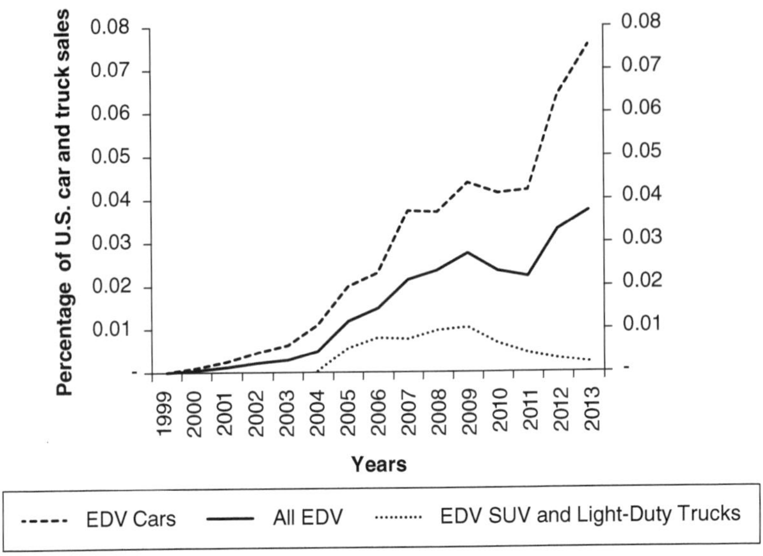

Figure 3.2 EDV market share, 1999 through 2012

Source: This figure combines the EDV sales data from Tables 3.1 through 3.3 with U.S. car sales data from Wards Auto, http://wardsauto.com/keydata/historical/UsaSa01summary.

by the sales of plug-in hybrid and all-electric vehicles using Li-ion batteries. Sales of these vehicles totaled 17,821 in 2011, 53,392 in 2012, and 96,602 in 2013 – accounting for 44, 42, and 56 percent of Li-ion sales in those years.

Among sport utility vehicles and light-duty trucks, sales of electric-drive vehicles peaked in 2009 at 1 percent and have since fallen to essentially nil. But electric drive cars have made steady headway, rising to slightly less than 4 percent of the U.S. car market in 2007, actually increasing their market share to just over 4 percent in 2011, and then exploding to 6.4 percent in 2012 and to 7.6 percent in 2013, as shown in Figure 3.2.

The data underlying Figure 3.1 and Figure 3.2 are provided in the following tables. Table 3.1 shows the dominance of the Toyota Prius. Sales of the Toyota Prius increased from 5,562 in 2000 to 223,905 in 2012.

Table 3.2 details the waxing and waning of the electric-drive sport utility vehicles and light-duty trucks. Total sales have decreased every year since 2006. The introduction of these vehicles in 2005 and their success in 2006 can be attributed in part to the Hybrid Motor Vehicle Credit (Section 1341) in the Energy Policy Act of 2005, which increased federal income tax credits for purchasing hybrid vehicles. The dependence of the amount of the tax credit on the relative improvement in the hybrid's fuel efficiency over that of a conventional vehicle of comparable size made the credit especially attractive to buyers of sport utilities and light-duty trucks. The credit was phased out for a given auto manufacturer

Table 3.1 NiMH HEV car sales, by model and year, 1999 through 2012

	1999	*2000*	*2001*	*2002*	*2003*	*2004*	*2005*	*2006*	*2007*	*2008*	*2009*	*2010*	*2011*	*2012*
Honda Insight	17	3,788	4,726	2,216	1,168	583	666	722	3	–	20,572	20,962	15,549	5,846
Toyota Prius[a]		5,562	15,556	20,119	24,627	53,991	107,897	106,971	181,221	158,886	139,682	140,928	136,463	223,905
Honda Civic[b]				13,707	21,771	26,013	25,864	31,253	32,575	31,297	15,119	7,336	4,703	–
Honda Accord[c]						653	16,826	5,598	3,405	198	1	–	–	–
Toyota Camry[d]								31,341	54,477	46,272	22,887	14,587	9,241	45,656
Lexus GS 450h[c]								1,784	1,645	678	469	305	282	607
Nissan Altima									8,388	8,819	9,357	6,710	3,236	103
Lexus LS 600h									937	980	258	129	84	54
Chevrolet Malibu										3,118	4,162	405	–	–
Saturn Aura										310	527	55	–	–
Fusion & Milan											17,022	22,232	11,286	14,100
Lexus HS 250h											6,699	10,663	2,864	649
Honda CR-z												5,249	11,330	4,192
Lincoln MKZ												1,192	5,739	6,067
Mercedes ML450												766	1	20
Mazda Tribute												655	484	90
Lexus CT 200h													14,381	17,671
Porsche Panamera S													52	570
Ford C-Max														10,935
Lexus ES														7,041
Toyota Avalon														747
Total	17	9,350	20,282	36,042	47,566	81,240	151,253	177,669	282,651	250,558	236,755	232,174	215,695	338,253

Sources: Sales data were compiled from J.D. Power, EDTA, Hybrid Dashboard, and Green Car Congress. See: www.hybridcars.com. See specifically, for 2011 and 2012, www.hybridcars.com/december-2012-dashboard.

Notes

[a] Includes all Prius models: Liftback, C, and V.

[b] The Civic hybrid sales are as reported by Honda through 2003 and 2004. Year 2005 and later data represent sales from EDTA, Hybrid Dashboard, and Green Car Congress.

[c] The Accord hybrid sales are from EDTA and Green Car Congress.

[d] Camry sales represent registrations from EDTA through 2006.

once it had, since December 31, 2005, sold 60,000 hybrids qualifying for the tax credit. The generosity of the tax incentives increased so considerably with the Energy Policy Act of 2005 that Beresteanu and Shanjun (2011) attribute 20 percent of hybrid sales in the United States in 2006 to tax incentives, compared with 5 percent of hybrid sales from 2001 to 2005.

Table 3.3 shows the introduction of the first dozen hybrid models with Li-ion batteries, dominated by the Hyundai Sonata, which was introduced in 2011. Sonata sales in 2011 and 2012 were about 20,000 per year. In comparison, sales of the Buick Lacrosse increased from about 1,800 in 2011 to 12,000 in 2012.

Table 3.4 details the introduction of plug-in hybrids and all-electric vehicles in 2011 and 2012. The GM Volt, Nissan Leaf, and Toyota Prius PHV are the dominant vehicles in this class.

The diffusion of NiMH and Li-ion technology through the market adoption of EDVs benefits both consumers and producers. Thus, from an economic evaluation perspective, the adoption of NiMH and Li-ion batteries generates consumer and producer surplus as well as broader-based societal benefits in terms of environmental health benefits associated with reduced vehicle emissions. These benefits stem principally from the improved fuel efficiency of EDVs compared with conventional ICVs; simply, less fuel is needed to travel a given distance.

Reduced fuel costs are an economic and energy benefit, and reduced environmental emissions are a health benefit to society; fuel costs and environmental emissions are reduced when ICVs are displaced by EDVs. A portion of these categorical benefits can be attributed to VTO's R&D investments to the extent that these investments have actually accelerated the adoption of NiMH and Li-ion battery technologies in HEVs. The attributable portion of these benefits is what drives the economic evaluation summarized in this book. These benefits are described and quantified in Chapters 4 and 5.

Table 3.2 NiMH HEV sport-utility and light-duty truck sales, by model and year, 2004 through 2012

	2004	*2005*	*2006*	*2007*	*2008*	*2009*	*2010*	*2011*	*2012*
Escape/Mariner[a]	2,993	15,960	22,549	25,108	19,522	16,480	12,088	10,089	1,441
Toyota Highlander[a]		17,989	31,485	22,052	19,391	11,086	7,456	4,549	5,921
Lexus RX 450h[a]		20,674	20,161	17,291	15,200	14,464	15,119	10,723	12,223
Saturn Vue[b]				3,969	3,399	2,656	50	–	–
Tahoe, Yukon, Escalade					7,612	7,192	3,857	1,936	1,801
Aspen & Durango					81	42	–	–	–
Silverado & Sierra						1,598	2,393	1,165	940
Porsche Cayenne							206	1,571	1,180
VW Touareg								390	250
Total	2,993	54,623	74,195	68,420	65,205	53,518	41,169	30,423	23,756

Sources: Sales data were compiled from J.D. Power, EDTA, Hybrid Dashboard, and Green Car Congress. See: www.hybridcars.com. See specifically, for 2011 and 2012, www.hybridcars.com/december-2012-dashboard.

Notes

[a] The Escape, Highlander, RX 400h, and GS 450h hybrid sales represent registrations from EDTA through 2006. The 2007 sales of Escape and GS450h are from Green Car Congress.

[b] The 2007 Vue hybrid sales are from EDTA (Jan–May only) and later sales are from Hybrid Dashboard and Green Car Congress.

Table 3.3 Li-ion HEV sales, by model and year, 2010 through 2012

	2010	*2011*	*2012*	*Total*
Mercedes S400	955	309	121	1,385
BMW ActiveHybrid	350	381	1,041	1,772
Hyundai Sonata		19,673	20,754	40,427
Buick Lacrosse		1,801	12,010	13,811
Kia Optima		403	10,084	10,487
Infiniti M		378	691	1,069
Buick Regal		123	2,564	2,687
Malibu		24	16,664	16,688
Honda Civic			7,156	7,156
Acura ILX			972	972
Audi Q5			270	270
Volkswagen Jetta			162	162
Year total	1,305	23,092	72,489	96,886

Sources: Sales data were compiled from J.D. Power, EDTA, Hybrid Dashboard, and Green Car Congress. See: www.hybridcars.com. See specifically for 2011 and 2012: www.hybridcars.com/december-2012-dashboard.

Table 3.4 PHV/EV (Li-ion) sales, by model and year, 2011 through 2012

	2011	*2012*	*Total*
GM Volt[a]	7,671	23,461	31,132
Nissan Leaf	9,674	9,819	19,466
Smart ED	388	139	527
Mitsubishi i	80	588	668
Ford Focus	8	685	693
Toyota Prius PHV[a]		12,750	12,750
Tesla Model S		2,400	2,400
Ford C-Max Energi PHV[a]		2,374	2,374
BMW Active E		671	671
Toyota RAV4 EV		192	192
Honda Fit EV		93	93
Year total	17,821	53,172	70,993

Sources: Sales data were compiled from J.D. Power, EDTA, Hybrid Dashboard, and Green Car Congress. See: www.hybridcars.com/december-2012-dashboard.

Note

[a] Plug-in hybrids.

Appendix 3.1: An overview of battery technology

Vehicle battery technology[2]

This appendix describes the application of energy storage technology to power light-duty passenger vehicles and trucks on the road. It begins with a general overview of the technology and then describes NiMH and Li-ion battery technologies and electrochemical capacitors. The purpose of this section is to review key performance characteristics, technical challenges that motivated public-sector investment, and concepts and terms necessary to evaluate VTO's investments.

Battery systems provide the power source for EDVs,[3] and the design of these systems determines the overall performance of an EDV in terms of how far it will travel (i.e., driving range), how quickly it will accelerate, how quickly it will be refilled (i.e., recharge rate), and how safe it will be in the event of an accident.

The battery system used in a typical EDV is a complex engineering structure consisting of a number of assemblies and subassemblies. These assembly components can be divided into four main categories: battery (e.g., integrated battery cells or modules), electrical controls (e.g., power electronics, electronic control units, battery management systems, and inductors), safety systems (e.g., crash sensor and current sensor), and maintenance facilities (e.g., blower and service plug).

The heart of an EDV battery system is a series of battery modules that are stacked in an array sized to meet the performance requirements of the EDV. Each module is composed of a series of battery cells and control electronics. If the battery system is the main energy source for vehicle propulsion, then it will be large and complex, and many battery modules (often 10 or more) will be needed to meet the performance requirements. In comparison, if the battery system is a supplement to another propulsion energy source (e.g., internal combustion engine), then the battery system will be significantly smaller and less complex than if the battery system were the main energy source. The number of cells in a module is determined by the performance specifications of the module.

Inside of each battery cell are three main components: an anode, a cathode, and an electrolyte. When a fully charged battery is connected to a motor or other electrical load, chemical energy stored in the battery causes a current to flow from the battery cathode through the external load and into the battery anode.

Because cell voltage and cell resistance are fundamental properties of the chemistry used in the battery cell, the choice of cell characteristics has an efficiency impact on the EDV battery system. In an ideal situation, the greatest amount of energy could be delivered if cell voltage was increased while simultaneously reducing cell resistance. Typically, however, it is not possible to achieve both high cell voltage and low cell resistance simultaneously. Thus, when designing a system, two cell options are available to use:

1 Cells with water-based electrolytes, including cell chemistries such as NiMH. Such cells typically have low cell resistance but limited cell voltage (typically below 1.5 volts).

2 Cells with electrolytes that are based on organic solvents (i.e., non-water based) such as Li-ion. Such cells typically have high cell resistance and high cell voltages (typically greater the 3.0 volts), compared with cells with water-based electrolytes.

Cost also plays a role in the choice of battery chemistry; water-based battery cells are typically available at a lower cost than Li-ion cells.

Typically, battery cells are made either by winding electrodes into a cylinder that can be inserted into a round cell – this is called a jelly roll configuration because of its resemblance to the food – or by stacking alternating plates of the anode and cathode – this is called a prismatic configuration.[4]

Because propelling a vehicle requires a large amount of energy, a single battery cell is insufficient to supply the energy needs. Instead, many batteries are connected to form a battery module and, depending on the configuration of the EDV, battery modules are connected to form a battery pack.

The cells that form the core of the battery system comprise different cell chemistries, each with certain advantages and certain limitations. Table A3.1 compares common cell chemistries. Lead acid batteries have been used for decades for starting, lighting, and ignition applications; and they offer high reliability at relatively low costs. However, lead acid batteries have a number of limitations, including very low energy density and a limited cycle life that precludes their viability for EDV applications. EDVs require more advanced battery technologies that can affordably achieve the power and energy density necessary to propel the vehicle for an extended range and that have the durability required to meet battery life requirements.

NiMH battery technology

NiMH batteries combine the chemistries of both NiCd[5] and NiH_2[6] to deliver cells that have high energy storage densities (both gravimetric and volumetric), long cycle life, and low cost. The principal difference between NiCd and NiMH cells is the replacement of the cadmium electrode in NiCd cells with metal hydride material. This replacement reduces cell weight and has a significant benefit for cell capacity. Because NiCd and NiMH cells share the same positive electrodes, the materials and manufacturing breakthroughs that have been developed over the many years of NiCd cell manufacturing can be directly applied to NiMH cells. This interoperability undoubtedly sped the commercial introduction of NiMH cells because only the negative electrode needed to be developed.

The reactions occurring at the NiMH battery electrodes are as follows:

$$NiO(OH) + H_2O + e^- \leftarrow\rightarrow Ni(OH)_2 + OH^- \text{ at the nickel positive electrode}$$

and

$$MH + OH^- \leftarrow\rightarrow H_2O + M + e^- \text{ at the metal hydride (M) negative electrode.}$$

Table A3.1 Technical performance of common cell chemistries used in EDV battery pack systems, circa 2010

Performance characteristics	*Lead acid*	*Nickel Cadmium (NiCd)*	*Nickel Metal Hydride (NiMH)*	*Lithium ion (Li-ion)*
Electrolyte	Water based	Water based	Water based	Organic
Nominal cell voltage (V)	2.0	1.2	1.2	3.6
Energy density (Wh/kg)[a]	35	40–60	60	120
Power density (W/kg)[b]	180	150	250–1,000	1,800
Cycle life[c]	4,500	2,000	2,000	3,500
Cost ($/kWh)[d,e]	$269	$280	$500–$1,000	Consumer electronics: $300–$800 Vehicles: $1,000–$2,000e
Battery characteristics	High reliability, low cost	Memory effect	Currently, best value and most popular for hybrid electric vehicles (HEVs)	Small size, light weight

Source: Lowe, Tokuoka, Trigg, and Gereffi (2010, p. 13).

Notes

[a]Chargeable electrochemical energy per weight of battery pack.

[b]Proportion of dischargeable electric energy to charged energy.

[c]The number of charging/discharging cycles in battery's entire life.

[d]Calculated exchange rate is $1 = 92.99 yen (05/14/2010). Ranges given are approximate.

[e]Liion batteries for consumer electronics have lower costs than those for use in vehicles because of high-volume production and a mature market.

This produces an overall cell reaction of:

$$NiO(OH) + MH \leftarrow\rightarrow Ni(OH)_2 + + M$$

The reactions occurring at the negative electrodes in NiH_2 and NiMH batteries are very similar, and the metal (M) hydride negative electrode has been the object of intense development in industry between 1992 and 2003. The material in this electrode must act as a hydrogen sponge and store large amounts of H_2, but it must also have high electrical conductivity and be able to oxidize the hydrogen gas. Breakthroughs in intermetallic hydrogen storage compounds for a variety of applications (e.g., fuel cells and solid-state refrigeration) have been leveraged in developing this technology.

Two broad classes of materials are used for the negative electrodes in NiMH batteries. The most common negative electrode material is the AB_5 chemistry in which A represents a rare earth element (usually lanthanum) and B represents one or more transition metals (typically a mixture of nickel, cobalt, or manganese). Another metal hydride chemistry that has also received extensive attention is the AB_2 class of hydrogen storage materials, in which A is typically titanium and B is typically nickel. The AB_2 chemistry was the first chemistry tested in NiMH batteries and offers the highest available theoretical capacity.[7] However, despite substantial research, this class of materials exhibits technically unacceptably high fade characteristics during multiple charge/discharge cycles; thus, its commercial potential is limited at this time. Consequently, virtually all of the commercial NiMH batteries on the EDV market contain the AB_5 chemistry, which has a more stable capacity even after 1,000 charge-discharge cycles.

Charging of NiMH cells reverses the electrochemical reactions shown in the equations above. Typically, recharging of NiMH cells occurs at a constant voltage between 1.4 and 1.6 volts per cell (reversed polarity from the discharge reaction), which is close to the typical operating voltage of NiMH cells (1.2 volts). Each battery cell is given a rate capacity value at the factory determined by the amount of energy that the battery can hold when new. This value is usually expressed in units of mAhr, and the value will decline or fade during cell usage. When a battery is charged at sufficient currents to completely recharge in 1 hour, this recharge rate is termed 1C rate. Likewise, if the battery is charged at twice this current, it is being charged at a 2C rate, and charging a battery at 1/100 of that current is charging at a C/100 rate.

A typical charging algorithm for NiMH cells consists of some combination of fast charge at high currents for short periods of time (typically less than an hour or < 1C) and trickle charge at much lower currents for an extended period of time (e.g., C/100).[8] If the charging current is low enough, trickle charging can be performed indefinitely on NiMH cells, producing cells that are completely charged when removed from the charger.

A critical component of charging NiMH cells is accurately determining when to terminate the cell charge. Failure to achieve proper fast charge cut-off can result in damage to the cell, including the possibility of an explosion. Among the most common charging termination methods for NiMH batteries are detecting the small cell voltage drop that occurs when the cell is fully charged and detecting the small increase in cell temperature that occurs when the cell is fully charged.

Fast charging is usually performed using a smart charger that adjusts the charging voltage and current to maximize the energy stored in the battery. Most cell manufacturers recommend that trickle charging commence upon the termination of fast charge to ensure that the cell is fully charged.

When cells are connected together, care must be exercised in how they are charged because the impedance of each cell will be slightly different.[9] Charging cells in a module requires that higher voltages be supplied to the module to charge each cell to the desired 1.4 to 1.6 volts per cell. In addition, the charging current to each cell must be adjusted to achieve an equivalent level of charge. The state of charge of each cell is monitored, often using the temperature change method, and fast charging is terminated when a portion of the cells becomes fully charged.

Li-ion battery technology

The Li-ion battery chemistry is a relatively new rechargeable battery technology that was first commercialized in the early 1990s. The Li-ion battery consists of a lithium-containing negative electrode and a positive electrode that is typically made from a metal oxide or a metal phosphate. The lithium ions are shuttled between the positive and negative electrodes. At the beginning of the cell discharge, the lithium is located in the negative electrode. During cell discharge, lithium ions are transported by the electrolyte to the positive electrode and reversibly inserted. During recharging, the process is reversed and lithium ions are removed from the positive electrode and inserted into the negative electrode. The highly reversible nature of the Li-ion shuttling process imparts excellent cycling performance to the Li-ion battery.

Table A3.2 summarizes several significant differences between Li-ion and NiMH batteries.

Li-ion batteries and NiMH batteries have some similarities; both are available in cylindrical and prismatic form factors, and cylindrical cells are often made in a jellyroll configuration. As a result, battery packs containing Li-ion cells can

Table A3.2 Selected differences between Li-ion and NiMH battery technology

Characteristic	*Li-ion*	*NiMH*
Electrochemical cycle	Lithium	Hydrogen
Electrolytes supported by	Organic solvents	Water
Cell voltage	Typically higher voltage per cell (nominally 3.5 volts or higher), allowing Li-ion batteries to store higher amounts of energy than is possible with lower cell voltages; higher cell voltage is enabled by using organic electrolytes that have higher voltage stability	Typically lower voltage per cell
Weight of materials	Intrinsically lighter weight materials (e.g., the density of carbon, a typical Li-ion battery electrode, is 2.3 grams per cubic centimeter, roughly a quarter the density of nickel)	Typically heavier materials (e.g., the density of nickel is nearly four times that of carbon), leading to lower energy storage capacity per unit weight (i.e., gravimetric energy density) compared with Li-ion
Safety	Organic solvents are flammable in combination with higher operational volts raise greater safety concerns compared with NiMH	Water-supported electrolytes are more stable than organic electrolytes

often replace those containing NiMH cells provided that an adjustment is made for the differences in cell voltages and charging requirements. Typically, Li-ion cells require one-third of the number of cells as NiMH because the cell voltage is roughly three times higher for Li-ion cells.

Originally, rechargeable lithium batteries were made with a lithium metal electrode. Although a lithium metal electrode exhibits the highest possible energy storage density, this electrode exhibits poor cycle life under repeated charge/discharge cycles. To overcome this limitation, commercial rechargeable Li-ion battery cells use carbon as the host material for the lithium in the negative electrode. Carbon has the ability to function effectively as a lithium sponge, reversibly absorbing and releasing lithium ions in the intercalation process. The chemical reaction occurring at the negative electrode is:

$$xLi^+ + xe^- + 6C \leftarrow\rightarrow Li_xC_6$$

Extensive studies have been performed on carbon materials for use as the negative electrode in Li-ion battery cells. These carbon materials can be broadly divided into amorphous materials (i.e., having no long-range structure) and those formed into crystalline carbon forms such as graphite. Amorphous carbons, also called hard carbons, have the advantage of the highest theoretical capacities and the best stability in common Li-ion battery electrolytes. This material was used in the first commercial Li-ion battery cells sold by Sony in 1991. Unfortunately, the large surface area of hard carbons leads to a high irreversible capacity (i.e., large capacity loss on the first cycle) and rapid decrease in cell capacity. Alternatively, graphitic carbons exhibit low irreversible capacities and little capacity loss with cycling (i.e., fading) and have become the dominant negative electrode material in Li-ion battery cells. However, graphitic carbons have a lower stability in certain electrolytes (e.g., propylene carbonate), and the material may undergo exfoliation under certain conditions rendering the electrode useless. Alternatively, mixes of hard and graphitic carbons are sometimes used as either coatings (e.g., hard carbon coating over graphitic particles) or mixed phase materials containing both graphitic and amorphous phases (e.g., mesocarbon microbeads). Although hard carbons were used in the first Li-ion battery cells, they have generally fallen out of favor for high duty cycle applications such as portable electronics and EDVs. In these applications, negative electrodes with a high percentage of graphitic content are used. Examples of these types of materials include graphitized mesocarbons, hard carbon-coated natural graphites, and synthetic graphites.

Recently, attention has shifted to other potential Li-ion battery negative electrode materials such as silicon and metallic tin, both of which exhibit the potential to serve as host materials for lithium. These materials can achieve considerably higher reversible charge capacities than carbon-based materials. However, they both suffer from large dimensional changes during the lithium insertion and removal process, and these changes result in poor mechanical stability and limited cycle life for electrodes made from these materials. An alternative negative electrode formulation is a partial replacement of carbon for nanosized metals of tin, silicon, or other elements to form a hybrid electrode.[10] This approach offers the potential to deliver higher capacities while retaining some of the mechanical support of the carbon structure.

Theoretically, lithium metal could also serve as the negative electrode in a lithium battery and deliver extremely high specific charge capacities of up to 3.9 Ahr/g. Lithium metal electrodes have performed exceptionally well in primary (i.e., non-rechargeable) cells. Unfortunately, the efficiency of a rechargeable lithium metal electrode is low, requiring an excess of lithium to achieve reasonable cycle life. In addition, lithium metal undergoes significant volumetric and morphological changes with continued cycling, and lithium metal is prone to form dendrites, which can short the two electrodes and produce an unsafe cell. For this reason, significant research is required before lithium metal electrodes displace carbon electrodes in rechargeable lithium batteries.

The positive electrode in a Li-ion battery cell is a metal oxide such as nickel oxide, cobalt oxide, or manganese oxide. These metal oxides also effectively function as lithium sponges and will absorb lithium during discharge and release it during recharge. The chemical reaction occurring at the cathode is illustrated for cobalt oxide as:

$$LiCoO_2 \leftarrow\rightarrow Li_{1-x}CO + xLi^+ + xe^-$$

Materials chosen for use as the positive electrode in Li-ion batteries must be able to absorb and release large amounts of lithium ions and must be chemically and structurally stable during operation. In selecting appropriate materials, the battery designer must strike the proper balance between power, energy, safety, durability and lifetime, and cost. A list of common metal oxides used in Li-ion battery chemistries is provided in Table A3.3. The most common chemistries are cobalt oxide (CoO_2) and nickel oxide (NiO_2) materials. Cobalt oxide is a relatively expensive material (i.e., poor cost performance), so combinations of CoO_2, NiO_2, and other metal oxides can often be used to improve cell cost without significantly affecting performance. Manganese oxide (MnO_2) and iron phosphate ($FePO_4$) are also commonly used because of their enhanced safety relative to the CoO_2 and NiO_2 materials. There is generally a trade-off between energy storage capacity and cathode safety, as shown in the table. The materials with the highest capacity (e.g., CoO_2 and NiO_2) can also present a safety concern if the battery pack is compromised. In contrast, materials such as iron phosphate and manganese spinel have better safety performance than other materials, but they also typically have a lower capacity.

Safe charging of Li-ion battery cells requires tight controls on the charging process to prevent (1) lithium metal plating on the negative electrode and (2) excess voltage (i.e., overvoltage) conditions that can produce unsafe thermal runaway reactions in the cell. Most Li-ion battery cells are charged to 4.2 volts (+/− 0.050 volts), although the maximum charging voltage will vary depending on positive electrode chemistry. Higher charging voltages, within a narrow range, will increase capacity but at the expense of cell lifetime. Charging at a significantly

Table A3.3 Illustrative snapshot of Li-ion chemistries with automotive applications, circa 2008

Chemistry	*Electrodes: positive (Negative)*	*Companies*	*Automotive status*	*Power*	*Energy*	*Safety*	*Life*	*Cost*
Lithium cobalt oxide (LCO)	$LiCoO_2$ (Graphite)		Limited applications	+	+	–	–	–
Lithium nickel, cobalt, and aluminum (NCA)	$Li(Ni_{0.85}Co_{0.1}Al_{0.05})O_2$ (Graphite)	JCI-Saft; GAIA; Matsuhita; Toyota	Pilot	+	+	+/–	+	+/–
Lithium iron phosphate (LFP)	$LiFePO_4$ (Graphite)	A123; Valence; GAIA	Pilot	+	+/–	+/–	+	+/–
Lithium nickel, manganese, and cobalt (NMC)	$Li(Ni_{0.33}Co_{0.33}Mn_{0.33})O_2$ (Graphite)	Litcel (Mitsubishi); Kokam; NEC Lamillion	Pilot	+/–	+/–	+/–	–	+/–
Lithium manganese spinel (LMS)	$LiMnO_2$ or $LiMn_2O_4$ ($Li_4Ti_5O_{12}$)	GS Yuasa; Litcel (Mitsubishi); NEC Lamillion; EnerDel	Development	+/–	–	+	+	+/–
Lithium titanium (LTO)	$LiMnO_2$ ($LiTiO_2$)	Altairnano; EnerDel	Development	–	–	+	+	–
Manganese nickel spinel (MNS)	$LiMn_{1.5}Ni_{0.5}O_4$ ($Li_4Ti_5O_{12}$)		Research	+	+/–	+	?	+/–
Manganese nickel (MN)	$Li_{1.2}Mn_{0.6}Ni_{0.2}O_2$ (Graphite)		Research	+	+	+	?	+/–

Source: Axsen, Burke, and Kurani (2008, p. 19).

Note: Simple + or – is used to indicate generally favorable or unfavorable. More details are provided in the original source.

higher voltage may produce an unstable cell that will fail catastrophically and in a potentially unsafe manner.

Typically, a Li-ion battery is charged in a two-step process: constant current followed by constant voltage. In the constant current portion of the charging cycling, a set current ranging from 0.5 to 0.8 of the rated charge capacity of the cell (i.e., 0.5C to 0.8C) is applied at an increasing voltage for approximately 1 hour. Then the cell is charged at a constant voltage (typically around 4.2 volts) with an exponentially decaying current until full charge is achieved.

Notes

1 Oakes (2006) traces the history of the battery from its most rudimentary origins in 250 BC (the "Baghdad Battery") through the development of the lead acid battery (familiar from its use in conventional vehicles) in the mid-19th century and up to and including the development in the 1970s of NiMH and Li-ion technology that is the basis for today's EDV batteries and the subject of this book.

2 This section draws from RTI International (2013), which benefitted directly from the contribution of Lynn Davis at RTI International.

3 Recall that the term EDV encompasses hybrids, plug-in hybrids, and all electric vehicles.

4 Cylindrical cells are made using the jelly roll process, whereas prismatic cells can be made using either the jelly roll or stacked plate structure.

5 Prior to the introduction of NiMH batteries, NiCd was the leading secondary battery technology for hand-held devices because of its low cost. Cells based on NiCd chemistry have been commercially available for more than 60 years, and a number of materials and manufacturing breakthroughs have been developed during this span to increase cell capacity and cell lifetime and reduce cost. Among these breakthroughs are the sintered nickel electrode to increase capacity, porous current collectors to increase capacity, and high volume manufacturing technologies to reduce costs. The energy stored by unit of mass (aka, gravimetric energy density) of the current generation of NiCd cells is typically only 40 to 60 Watt hours per kilogram (Wh/kg), while the energy stored per unit volume (aka, volumetric energy density) is typically between 50 and 150 Watt hours per liter (Wh/L).

Despite the advantages of the NiCd cell, two significant limitations of this chemistry are the energy density and the "memory effect." The energy density of NiCd cells is limited by the weight of the materials used to construct the cell. If NiCd batteries are charged and discharged to the same level on a consistent basis, the cell develops a basis against operating outside this range, thus effectively reducing battery capacity. This is the basis of the "memory effect" that occurs with NiCd batteries.

6 The NiH_2 battery is a high energy density secondary battery that is widely used in satellites and other aerospace applications. The battery consists of the nickel positive electrode, a platinum-impregnated carbon black negative electrode, and a potassium hydroxide electrolyte. Hydrogen is supplied to the cell in the form of gaseous hydrogen contained in a pressurized vessel containing the cell components. The key performance advantages of this battery are its high gravimetric energy density (75 Wh/kg) and its long cycle life (>20,000 cycles at 80% depth of discharge). However, the cost of this battery is relatively high so it is used in specialized applications such as aerospace.

7 Because of its potentially large theoretical capacity, Energy Conversion Devices (ECD) of Auburn Hills, Michigan, and its battery subsidiary, Ovonics, invested heavily in the AB_2 chemistry.

8 Charging a battery is analogous to filling a bucket with water. A large hose can be used to fill the bucket quickly, but the water flowing to the bucket must be turned off early to avoid spillage or other problems. In contrast, a smaller hose or even an eye dropper can be used to fill the bucket exactly to the desired level without any spillage. However, such procedures would significantly increase the time required to fill the bucket. For batteries, a large current can be applied at a negative voltage that is slightly above the discharge voltage (i.e., 1C rate or higher). However, this process, which is termed fast charging, will not fill the battery fully without causing some potential long-term problems. A slower charging procedure, termed trickle charging because a trickle of electrical current is used for charging, can charge cells very accurately, but takes a long time. Usually, a combination of fast charge and trickle charge is used in recharging batteries.

9 This charging is analogous to simultaneously filling buckets of slightly different sizes and apertures from a single hose.

10 See Novák et al. (2010).

4 Measurement of economic and energy benefits

The measurement of economic and energy benefits attributable to the U.S. Department of Energy's R&D investments in battery technologies comes from extensive interviews with informed scientists and technicians throughout the United States. The interview guide is presented, discussed, and annotated in this chapter, as are the methods used to translate informed expert responses into monetized economic and energy benefits data. Those monetized data are also presented.

Construction of an interview guide

Background information related to an initial understanding of the measurement of economic and energy benefits was obtained through unstructured telephone interviews with several key individuals at USABC and at several U.S. automobile companies. Based on these background interviews, there appeared to be areas of general agreement on a number of issues related to the method for identifying and measuring the economic and energy benefits attributable to VTO's R&D investments.

Noteworthy from these informational interviews are two important observations, important in the sense that they influenced both the construction of the interview guide from which economic and energy benefits were ultimately identified from industry and other experts, and important for a meaningful interpretation of what would be interviewees' responses.

There was general agreement that VTO's R&D investments accelerated both the transition from conventional ICVs to EDVs and the adoption of EDVs in the market. More specifically, those with whom we spoke for background information acknowledged that there were indeed important technical improvements in battery technologies, particularly NiMH and Li-ion battery technologies, that were a direct result of (i.e., that were directly attributable to) VTO's R&D investments. However, these individuals were also emphatic that the main economic benefits associated with VTO's R&D investments were associated with the increase in the rate of market adoption of EDVs. The consensus opinion was that VTO's R&D investments, and the subsequent technical improvements in NiMH and Li-ion battery technologies, increased the rate of adoption of EDVs on the road by 5 to 10 years. Information from these background interviews is summarized in Box 4.1.

Box 4.1 Consensus opinions from individuals interviewed at USABC and at U.S. automobile companies

- VTO's R&D investments in energy storage technologies accelerated the transitional period from the use of lead acid battery technology to alternative battery technology and the resulting adoption of EDVs on the road by 5 to 10 years.
- Universities were not involved in NiMH battery research in the 1990s and afterwards; NiMH battery research was industry driven.
- Absent VTO's R&D investments, the U.S. automotive industry would today likely be importing all of its Li-ion batteries.
- There has been a significant lag in time between VTO's R&D investments and the realization of that technology as an innovation used in EDVs.
- The major economic impact of VTO's R&D investments is primarily realized through the increase in the market adoption of EDVs.

Note: The VTO provided contact information for seven individuals who were either associated with USABC or were/are at U.S. car companies. Three of these seven individuals were willing to provide background insight on general issues related to identifying and measuring the economic benefits attributable to VTO's investments in NiMH and Li-ion battery technologies: Ted Miller, Senior Manager of Energy Storage and Materials and Energy Research at Ford and member of the Management Committee of USABC; Gary Henriksen, Senior Electrochemical Engineer and head of battery research at ANL; and Mark Verbrugge, Director of Chemistry and Materials R&D at General Motors. The comments in this box are consensus opinions, and no specific comments should be attributed to any specific individual who was interviewed.

Information gleaned through these background interviews informed the preparation of the interview guide. The guide was used as a pedagogical device to focus the telephone interview discussions and to ensure that the data collected accurately represented the economic and energy impacts attributable to VTO's R&D investments. The interview guide appended to this chapter has been annotated with paraphrased representative interviewee responses.

Our reliance on information from these background interviews for the preparation of the interview guide was premised on the background and experience of those with whom we spoke. As mentioned in the note to Box 4.1, these are experienced individuals who have a broad understanding of the supply chain for battery technology as well as the impact of technical improvements in battery technology on EDV market activity.

It was critical for the data collection process that the interview guide transitioned the interviewee from VTO's R&D investments through resulting technical improvements in battery technologies to the market adoption of EDVs.

Understanding the impact of VTO's R&D investments on the market adoption of EDVs is key to quantifying precisely and accurately relevant economic and energy, and environmental health, benefits.

Prior to each telephone interview, each industry scientist who accepted the invitation to participate in this study was sent electronically a copy of the interview guide. Along with the interview guide, background information on the purpose and scope of this evaluation study was provided as well as an emphasis on the fact that the interview guide was literally a guide or roadmap to the discussion about the economic impacts associated with VTO's R&D investments.

Scientists at university and national laboratories were also sent the interview guide, but each was asked to complete it as a survey instrument and return it electronically; in rare cases were university or government laboratory scientists interviewed by telephone.

Questions 1 and 2 on the interview guide ask respondents to describe current and previous R&D projects related to NiMH and Li-ion technologies and to explain how those projects had been affected by VTO's investments. Responses to these questions provided background information that complemented what was learned from the initial informational interviews summarized in Box 4.1. Questions 1 and 2 in our guide are especially important. Responses to these two questions were assuring that each individual interviewed not only was familiar with VTO's R&D investments, but also that he/she was familiar with NiMH and Li-ion battery technology.[1]

The remaining questions on the interview guide ask respondents to consider a counterfactual situation in which VTO had not supported NiMH and Li-ion battery technologies through its R&D investments. Specifically, Question 3 asks how industry-wide battery research efforts and outcomes would have been different without VTO's funding support. Again, responses to this background question offered assurance about the knowledge base of each respondent.

Question 4 asks how the advancement in the state-of-the-art battery technology over roughly the past fifteen years would have been different without VTO support. Respondents were presented with figures showing the advancement of NiMH and Li-ion battery technologies in three dimensions: battery life, gravimetric energy density, and cost. Discussions during the interviews about these technical dimensions were intended to be used as background information to characterize the battery technology; responses to questions about these technical dimensions were not intended to be used to quantify the attributable impacts of VTO's R&D investments.

Respondents were then asked to describe the counterfactual situation without VTO support and to confirm their verbal descriptions to us by using a graphical device within the interview guide.[2] All of the stylized base figures that were provided to the interviewees in the interview guide were derived from DOE sources, EERE merit review presentations, and International Electronics Manufacturing Initiative (iNEMI) roadmaps.[3] Thus, each interviewee was provided with an informed description of the current state-of-the-art for NiMH and Li-ion battery technologies. Each interviewee's verbal and graphical responses to Question 4 were useful regarding how best for us to transition meaningfully to Question 5.

Question 5 asks how the market adoption of NiMH and Li-ion battery technologies through the purchase of EDVs in the United States would have been different in the counterfactual situation of no VTO R&D support. Responses to this question confirmed the construct validity of the survey guide as well as the probative value of the counterfactual approach initially confirmed through the background interviews; see Box 4.1. Interviewees were graphically presented with factual information describing the current market share of EDVs in the United States from 1999 to 2011. Each was asked to describe the counterfactual situation about what the market adoption of EDVs would have been in the absence of VTO's R&D support. Care was taken during this discussion to emphasize to each interviewee that a number of factors could influence the current market share of EDVs, and that the counterfactual aspect of Question 5 only considered the specific impact of VTO's R&D support. Such factors included the global acceptance of EDVs, technical advancements by Japanese battery manufacturers, and other U.S. regulations such as increased Corporate Average Fuel Economy (CAFE) standards and tax credits for the purchase of EDVs. Then, after a careful discussion of the counterfactual issue that was relevant to the study, each interviewee was asked to confirm his/her description graphically.

The market adoption of EDVs over time is certainly related to changes in both the supply and the demand of EDVs, and certainly separating supply effects from demand effects is important from an economic perspective, albeit that the separation of supply effects from demand effects is complicated. However, for the purpose of this study, that is for quantifying the market adoption of EDVs over time in the absence of VTO's R&D investments, of interest is only the resulting increase in EDVs actually on the road.

Imbedded in the interview discussion on Question 5 is the important issue of attribution. To ensure that each interviewee was indeed offering an opinion about the counterfactual market adoption of EDVs in the absence of VTO's R&D investments, each was asked for clarification of this focused structure in Question 7. As a transition to Question 7, Question 6 asks about the share of the EDV market in the United States with Li-ion battery technology in the absence of VTO's R&D support. Expectations were that the interviewees would be relatively more familiar with the impact of Li-ion battery technology than NiMH battery technology, and thus this question segues to Question 7.

Question 7 was designed to confirm that each interviewee did indeed understand the nature of our counterfactual inquiry because his/her responses to Question 5 would be the basis for the quantification of economic and energy, and environmental health, benefits. Nearly 90 percent of those interviewed responded that they fully understood the nature of the counterfactual inquiry. Only five interviews paused and responded that part of their graphical response to Question 5 did include other economic impacts such as the global acceptance of EDVs and technical advancements by Japanese battery manufacturers. Based on further discussion, the graphical shift in the market adoption curve from Question 5 was adjusted. As discussed below, the adjustment factor that was imposed on the graphical response from these five scientists ranged between 10 and 20 percent based on interview information.

Sample of interviewees and respondents

The sampling population was identified from information provided by VTO and from independent research on three sources of experts: scientists in companies funded by VTO for either NiMH or Li-ion battery research, university scientists funded by VTO for either NiMH or Li-ion battery research, and national laboratory scientists funded by VTO for either NiMH or Li-ion battery research. Specifically, two overlapping sources were used for identifying these three samples: background information from VTO personnel on companies that VTO had funded and annual reports published by VTO.

The representativeness – in terms of receipt of DOE R&D and in terms of stages in the battery value chain – of those who were interviewed and thus from whom relevant evaluation information was obtained is proffered based on the following information.

The funded organizations (companies, universities, and national laboratories) considered for the sampling population totaled 148. Column 2 of Table 4.1 shows the number of funded organizations, by category, as reported by VTO in its annual reports. At least one contact individual at each organization was identified either directly on the basis of information provided by VTO, or indirectly through VTO annual reports (e.g., university scientists), or through our independent research. As shown in column 3, contact information was obtained for scientists at 95 of the 148 organizations, or 64 percent of the sampling population. Contact information was obtained for at least one scientist at all of the university and national laboratories but only for at least one scientist at 49 percent of the funded companies. Fifty four (see column 5) scientists from 40 (see column 4) of the 95 organizations for which there was contact information agreed to participate in the study directly by telephone or indirectly by a personalized electronic survey (e.g., university scientists).[4] These 54 scientists from 40 organizations represent an average organization coverage ratio of 42 percent (see column 4).

The organizational group most heavily represented in the sample was funded companies (n=25). Scientists from these companies were either current or former R&D managers, thus each was able to provide through his/her responses a broad perspective about the EDV market.

Collectively, these 25 companies were the recipients of 98 percent of DOE's R&D investments in NiMH battery technology and 76 percent of DOE's R&D investments in Li-ion battery technology. This R&D funding came through USABC contracts between 1995 and 2010, based on USABC contract funding summaries provided by DOE; see Table 4.2.

DOE provided R&D funding to 41 companies in five major segments along the Li-ion value chain as characterized by Figure 4.1. Among the 25 company respondents are 22 of those 41 companies (see column 2 in Table 4.3): 3 of 7 key

Table 4.1 Participants in the data collection process, by stakeholder category

(1) Category of organization	*(2) Funded by DOE/ USABC since 1992*	*(3) With contact information (percentage of funded organizations)*	*(4) Organizations interviewed/surveyed (percentage of contacted organizations)*	*(5) Number of individuals participating*
Companies	104	51 (49%)	25 (49%)	25
Universities	28	28 (100%)	5 (18%)	6
National laboratories	16	16 (100%)	10 (63%)	23
Total	148	95 (64%)	40 (42%)	54

Table 4.2 Distribution of company participants by battery technology area, n=25

(1) Battery technology area	*(2) DOE R&D investment in USABC contracts, 1995–2010 ($millions)*	*(3) R&D investment amount represented in the sample ($millions)*	*(4) Percentage of USABC contract R&D investment represented in the sample*
NiMH	29.8	29.2	98%
Li-ion (including lithium polymer)	157.6	119.0	76%

Source: Based on USABC Contract Summary 1995–2010, provided by DOE.

materials manufacturers, 6 of 13 cell components and electronics manufacturers, 7 of 9 integrated systems manufacturers, 2 of 2 original equipment manufacturers (OEMs), and 4 of 10 U.S. venture capital startups.

A list of the companies, universities, and national laboratories that participated in the survey follows the interview guide in Table A4.1 of the appendix to this chapter.

Counterfactual battery life, energy density, and cost improvement without VTO support

For descriptive purposes, interviewees were asked in Question 3 of the interview guide how industry-wide battery research efforts and outcomes would have been different in the absence of VTO's support. The mean response was that VTO accelerated industry-wide R&D investments by just over six years, with a standard deviation (std. dev.) of less than one year (see Table 4.4; mean values are in column 1, the standard deviations are in column 2, and the sample size is in column 3.[5])

Key materials →		Cell components and electronics →		Integrated systems →	Vehicles
Cathode precursor	***Cathode***	***Other cell components***	***Electronics***	***Integrated systems***	***Vehicles***
Lithium compounds	Active material	Package – steel or aluminum can – Laminate film	Mechanical components	Major Li-Ion battery cell/pack players	Relevant automotive OEMs
Co compounds	Al Foil	Lead	Electrical components		Additional relevant OEMs
Mn compounds	Polymer binder (PVDF)	Insulator	Electric components		
Ni compounds	Carbon electric conductor	Safety vent			
		Gasket			
Anodes	***Anode***	PTC			
Purified natural graphite	Active material (graphite)	Center pin			
Graphite precursor	Cu Foil	Tab			
	Binder				
	Carbon electric conductor				
Electrolyte	***Electrolyte***				
Organic solution DMC/MC/EC/MEC					
Li-Salt (LiPF6)	***Separator***				
Polymer precursor for polymer battery					

Figure 4.1 Value chain of Li-ion batteries for vehicles

Source: Based on Lowe et al. (2010, p. 33) and the Center on Globalization, Governance and Competitiveness at Duke University, Durham, NC.

Table 4.3 Distribution of evaluation participants along the Li-ion battery value chain

(1) *Value chain segments*	*(2)* *Companies receiving VTO funding for Li-ion research*	*(3)* *VTO-funded companies interviewed*	*(4)* *Coverage ratio by value chain location*
Key materials	7	3	43%
Cell components and electronics	13	6	46%
Integrated systems	9	7	78%
Vehicle OEMs	2	2	100%
U.S. venture capital startups	10	4	40%
Total	41	22	54%

Source: Based on the value chain segments in Lowe et al. (2010, p. 34).

Note: OEM refers to original equipment manufacturers.

Table 4.4 Battery life, energy density, cost, and Li-ion EDV sales improvement attributable to VTO R&D investments

Metric	*(1)* *Mean*	*(2)* *Std. Dev.*	*(3)* *n*
Question 3:			
Number of years by which industry-wide R&D was accelerated by VTO	6.1 years	0.8	15
Question 4:			
Percentage of progress from 1998 to 2012 attributable to VTO			
Battery life	35%	3%	38
Energy density – NiMH	33%	12%	12
Cost – NiMH	30%	8%	12
Energy density – Li-ion	33%	5%	37
Cost – Li-ion	33%	4%	37
Question 6:			
Percentage of Li-ion share of EDV market attributable to VTO	66%	7%	21

Note: The conventional estimate of the standard deviation of the sample mean, calculated by dividing the standard deviation of the sample of n answers by the square root of n, is reported.

Also shown in Table 4.4 are mean responses to Question 4. Specifically, interviewees were asked, again for descriptive purposes, to describe graphically advancements in the state-of-the-art characteristics of NiMH and Li-ion batteries since 1998. From the responses to these questions, the mean opinion that battery life has improved by 35 percent, energy density for both NiMH and Li-ion batteries has improved by 33 percent, and the cost of NiMH batteries has decreased by 30 percent and that of Li-ion batteries by 33 percent; see column 1 in Table 4.4.

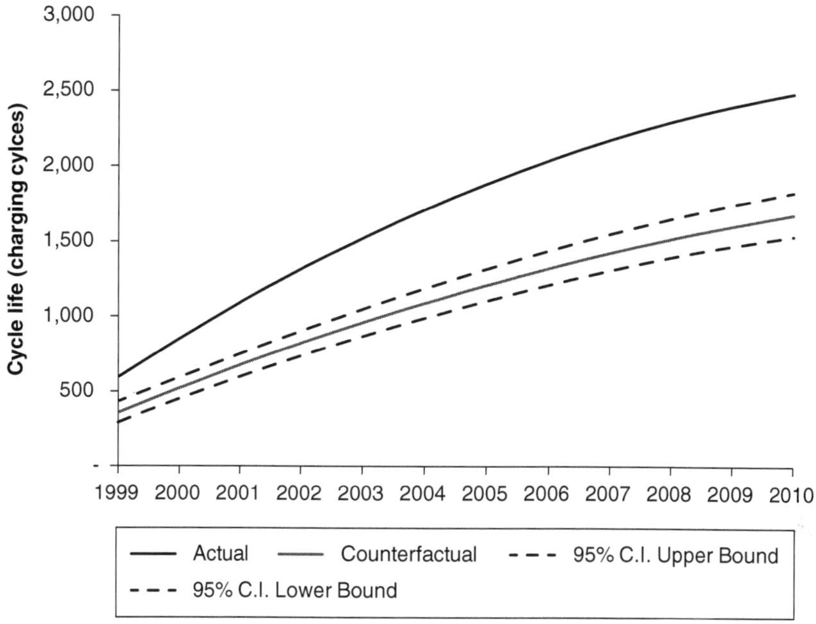

Figure 4.2 Counterfactual battery life (charging cycles) improvement without VTO support

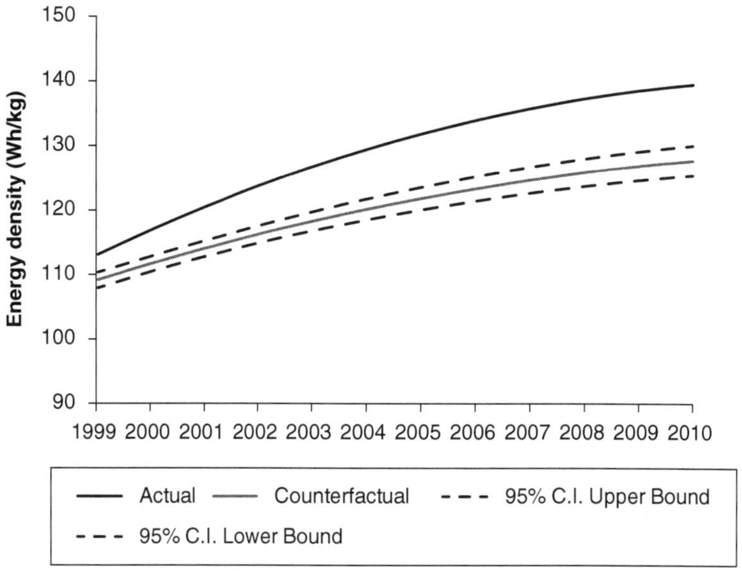

Figure 4.3 Counterfactual energy density (Wh/kg) improvement without VTO support

Figures 4.2 through 4.4 illustrate the interviewees' responses to Question 4 for Li-ion battery technology.

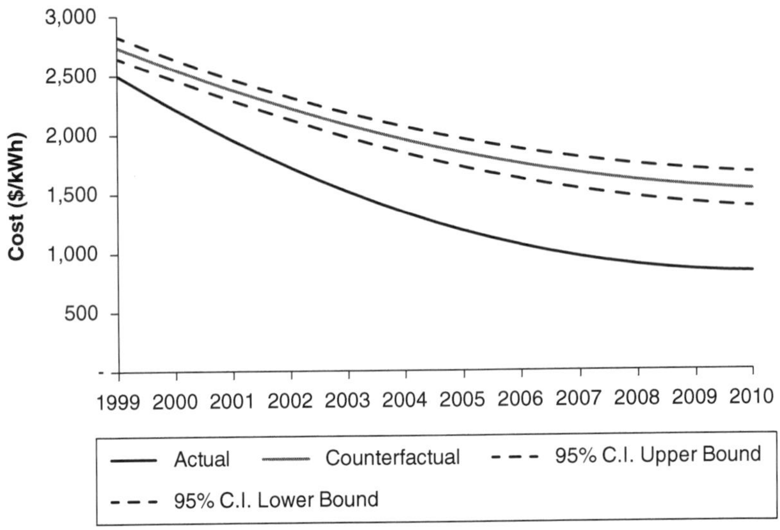

Figure 4.4 Counterfactual cost ($/kWh) improvement without VTO support

Question 6 asked respondents if the Li-ion (as opposed to NiMH) share of the EDV market would have been different without VTO R&D investments, and if so, how. Out of 37 respondents providing any kind of response, 4 said there would be no Li-ion batteries on the road without the VTO's investments and 8 more said that the Li-ion share of the EDV market would be dramatically or significantly less. One respondent said that the Li-ion share would be the same. The estimate that 66 percent of the Li-ion share of the EDV market can be attributed to VTO investments was obtained by averaging the 21 quantitative responses to this question; see column 1 in Table 4.4.

Counterfactual EDV adoption in the absence of VTO support

Two options were considered for collecting relevant market information on the impact of VTO's R&D investments in battery energy storage technology. The first option that was considered was to ask interviewees about the technical improvements in NiMH and Li-ion battery technologies that resulted from VTO's R&D investments, and then to use third party information to translate technical improvements to activity in the EDV market. The second option, and the one that was chosen for this study, was to rely on the expertise of interviewees to address the impact of VTO's R&D investments in NiMH and Li-ion battery technologies on the market adoption of EDVs. This approach not only followed the advice of those interviewed during the background information stage of the study (see Box 4.1), but also it is the logical approach given the expertise of the respondents.

Thus, the evaluation metrics below are quantified using the interview information primarily from Question 5 (and in 5 instances, as was noted above, the interview information from Question 5 weighted by interview information from Question 7).

Interviewees were asked in Question 5 to consider the market adoption of EDVs in the United States since 1999, as shown in Figure 4.6, under a counterfactual situation of no VTO R&D investments for either NiMH or Li-ion battery technologies. Interviewees noted that their perspectives touched not only on counterfactual performance, but also on whether the intended vehicle itself was viable as an EDV given the counterfactual performance without VTO support. Interviewees described their opinion of the counterfactual and confirmed that opinion using the interview guide's schematics.[6] Thus, we were able to determine, by year, the percentage of EDV sales attributable specifically and uniquely to VTO's R&D investments.

The values in Table 4.6 summarize the interviewees' responses to Question 5. They were calculated using a four-step approach as described below.

Step 1

Each respondent was presented with an adoption curve based on Figure 4.5. The picture actually presented to respondents is provided with the interview guide in the appendix (Figure A4.4).

Table 4.5 Percentage of market adoptions of EDVs attributable to VTO R&D investments in NiMH and Li-ion battery technologies, n=44

Year	*(1)* Mean	*(2)* *Std. Dev.*
1999	40	6
2000	44	5
2001	47	5
2002	49	4
2003	51	4
2004	52	4
2005	51	4
2006	50	4
2007	47	4
2008	44	4
2009	41	4
2010	38	4
2011	36	4
2012	34	4

Note: The conventional estimate of the standard deviation of the sample mean, calculated by dividing the standard deviation of the sample of n answers by the square root of n, is reported.

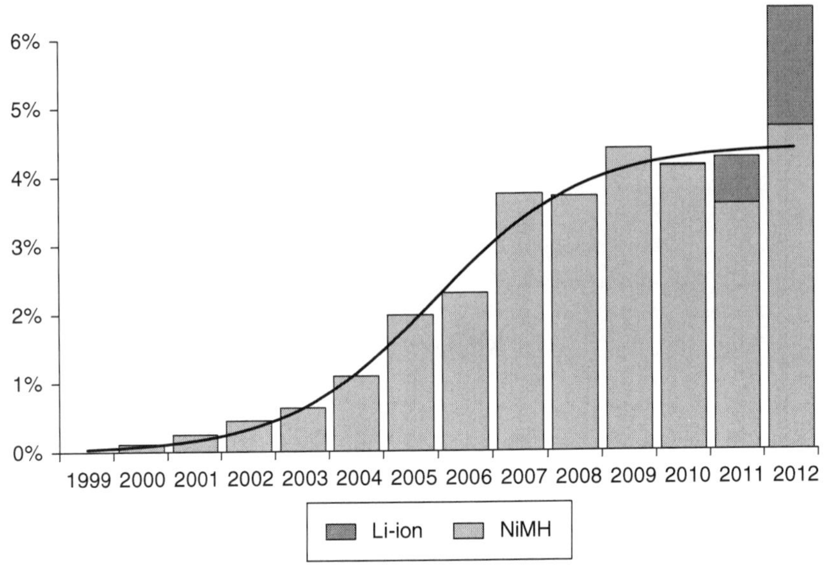

Figure 4.5 Market adoption of EDVs in the United States since 1999; percentage of cars sold in the United States powered by NiMH or Li-ion battery technology

Note: The stylized adoption curve was constructed before data were available for 2012 and thus does not reflect the marked increase in the market share of (especially Li-ion) EDVs in 2012, which continued in 2013 (see Figure 3.2).

The general formula for a sigmoidal curve, like the one in Figure 4.5, is:

$$P_t = M / \left\{ 1 + e^{\left(\frac{t^{mid} - t}{a} \right)} \right\} \tag{4.1}$$

where e is the base to the natural logarithms, M is the maximum attained by the sigmoidal curve, t is analytic time (in our case 1999 corresponds to $t = 0$), t^{mid} is the number of years taken by the curve to climb half of the way to M, and a is an index of how gradually the sigmoidal curve rises. The units of M were chosen to be the percentage market share of EDVs (rather than the approximate number of EDVs sold), so P_t is likewise the percentage market share of EDVs at time t.

Step 2

Each respondent's (respondent i's) description and opinion were characterized in terms of the values M_i, t_i^{mid}, and a_i, chosen to most closely replicate respondent i's redrawn curve.

Five of the 44 interviewees answered Question 7 – *For clarification, are the differences that you have described between the actual and counterfactual scenarios due entirely to VTO's financial support of NiMH and Li-ion technologies* – with a value of less than 100 percent, as noted above. These 5 coded responses were recorded as weighted averages: $w_i M_i + (1 - w_i)M_0$, where w_i is the respondent's answer to Question 7 (as a decimal between 0 and 1), M_i is the parameter of the counterfactual curve, and M_0 is the parameter of the original curve shown to the respondent. The same adjustment was performed on t_i^{mid} and a_i.

To reiterate, when an interviewee responded to Question 7 with a value less than 100 percent, that individual's response to Question 5 was adjusted or weighted to ensure that the values in Table 4.5 are solely, in the opinion of the interviewee, attributable to VTO's R&D investments.[7]

Step 3

Based on each respondent's response to Question 5 and his/her response to Question 7, the parameterized values for M_i, t_i^{mid}, and a_i were translated into values for the years 1999 through 2012 ($P_0^i, P_1^i, \ldots, P_{13}^i$) and compared with the values associated with the adoption curve originally shown in Question 5. From this comparison, the percentage of EDV sales attributable to VTO support in each year (A_t^i is the attributable percentage of EDV sales at time t, based on respondent i's answer) is:

$$A_t^i = 100\left(1 - \frac{P_t^i}{P_t}\right) \tag{4.2}$$

Step 4

For each year t, the attributable percentage of market adoption of EDVs was averaged across the 44 (of 54) respondents who provided an answer to Question 5. That is:

$$\overline{A_t} = \sum_{i=1}^{n} A_t^i / n \tag{4.3}$$

This average of $\overline{A_t}$ from the equation above is reported for each year in Table 4.5. Figure 4.6 shows the actual and average counterfactual adoption curves with the 95 percent confidence interval for the counterfactual.

The mean values in column 1 of Table 4.5 are used to determine fuel savings associated with the economic impact of VTO's R&D investments on the market adoption of EDVs – HEVs, PHEVs, and EVs – as discussed below.

The variation of opinions among informed scientists is reflected in the confidence interval in Figure 4.6. Not only is this variation of opinion expected in an evaluation in which the economic consequences of public moneys are manifested

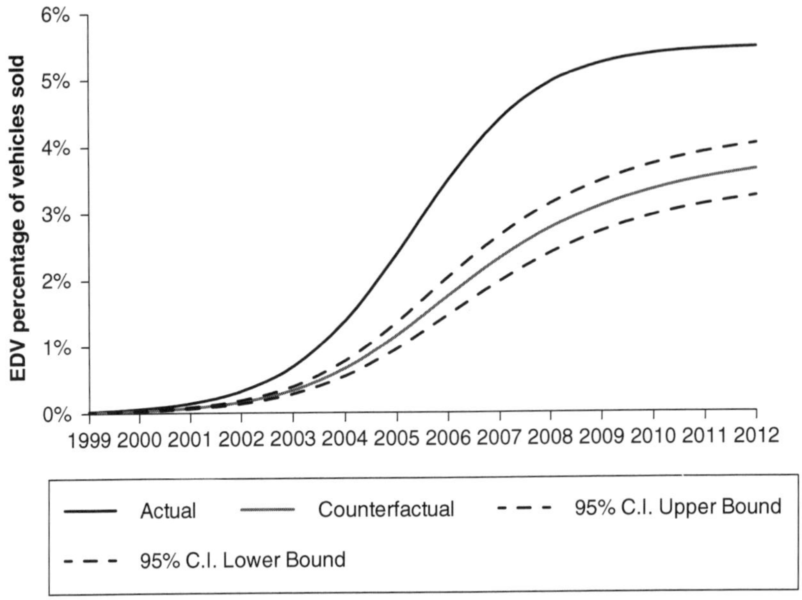

Figure 4.6 95 percent confidence interval on percentage of market adoption of EDVs attributable to VTO's R&D investments (actual adoption curve comes from Figure 4.5)

across an integrated supply chain (see Box 4.1), but also such variation gives credibility to the construction of the sampling population and interview sample. Absent such variation, selectivity might bias the evaluation metrics even through means from the interview sample are used.

Vehicle miles and fuel savings attributable to VTO's R&D investments

The estimation of economic and energy benefits attributable to VTO's R&D investments in NiMH and Li-ion battery technologies is based on the fuel savings associated with adopting EDVs. Fuel savings attributable to VTO's R&D investments are measured using a three-step disaggregated approach as described below.

Step 1

The market adoption of HEV, PHEV, and EVs attributable to VTO's R&D investments was calculated. These values are in columns 5, 6, and 7 of Table 4.6.

As an illustrative example of the calculations in Table 4.6, consider the year 1999. Of the 17 vehicles sold in 1999 (column 1), 39.75 percent (40 percent or 0.40 in column 4) were attributed to VTO. Thus, 17 x 0.3975 = 6.7575 attributable vehicles, rounded to 7 vehicles in column 5.

Table 4.6 Market adoption of HEV, PHEV, and EVs in the United States attributable to VTO's R&D investments

Year	(1) HEV sales	(2) PHEV sales	(3) EV sales	(4) Percentage of sales attributable to VTO (rounded)	(5) HEV sales attributable to VTO (rounded)	(6) PHEV sales attributable to VTO (rounded)	(7) EV sales attributable to VTO (rounded)
1999	17			40	7		
2000	9,350			44	4,097		
2001	20,282			47	9,519		
2002	36,042			49	17,748		
2003	47,566			51	24,161		
2004	84,233			52	43,396		
2005	205,876			51	105,505		
2006	251,864			50	125,536		
2007	351,071			47	166,371		
2008	315,763			44	139,782		
2009	290,273			41	118,982		
2010	274,648			38	104,338		
2011	269,210	7,671	10,150	36	95,621	2,725	3,605
2012	434,498	38,585	14,587	34	146,021	12,967	4,902

Note: The EDV sales by year in columns 1 through 3 come from Tables 3.1 through 3.3. Column 2 comes from Table 3.3 and includes sales of the Volt, Prius PHV, and C-Max Energi PHV. The percentage of sales attributable to VTO was reported in column 1 of Table 4.5. Actual sales of each EDV type were multiplied by this average percentage to obtain the attributable sales values in columns 5 through 7 of Table 4.6. These constructed attributable sales values were then carried over to Step 2.

Step 2

The number of attributable HEVs, PHEVs, and EVs on the road in a given year were calculated from the number of attributable vehicles sold (see Table 4.6) along with the assumption that a vehicle stays on the road for 11 years and then is scrapped. Support for this assumption comes from the publicly available data on the miles driven for vehicles of a specific age; see Table 4.7.[8]

Based on the assumption of a vehicle being scrapped after it has been 11 years on the road (i.e., one year after it has turned 10 years of age), Table 4.8 was constructed using the sales of HEVs attributable to VTO in column 5 of Table 4.6. Note that in Table 4.8, the 7 attributable HEVs purchased in 1999 continue to be on the road through 2009.

Tables 4.9 and 4.10 are similarly constructed from columns 6 and 7 of Table 4.6 and show the number of attributable PHEVs and EVs on the road, by year and by vehicle age. The 12,967 new PHEVs from column 6 in Table 4.6 that were first put on the road in 2012, for example, remain on the road through 2022.

Step 3

Table 4.11 shows the calculated fuel savings associated with the attributable EDVs purchased from 1999 through 2012. Column 1 reports the sum of all columns for the given year from Table 4.8. Similarly, columns 2 and 3 report the sum of all columns for the given year from Tables 4.9 and 4.10, respectively. Note that columns 1, 2, and 3 are not directly used in any subsequent calculations, because they sum together vehicles of different ages that are assumed to be driven different amounts; see Table 4.7.

Table 4.7 Average miles driven by vehicle age

(1) *Vehicle age*	*(2)* *Average annual miles*
Under 1	14,350
1	14,450
2	13,950
3	12,900
4	12,550
5	12,400
6	11,950
7	11,750
8	11,200
9	11,050
10	8,350

Source: Based on U.S. DOE (2012, Table 8.10).

Note: The average annual miles driven, by vehicle age, is the average of 2001 and 2009 self-reported miles driven.

Table 4.8 Attributable HEVs on the road, by year and vehicle age

Year	*<1 year*	*1 year*	*2 years*	*3 years*	*4 years*	*5 years*	*6 years*	*7 years*	*8 years*	*9 years*	*10 years*
1999	7										
2000	4,097	7									
2001	9,519	4,097	7								
2002	17,748	9,519	4,097	7							
2003	24,161	17,748	9,519	4,097	7						
2004	43,396	24,161	17,748	9,519	4,097	7					
2005	105,505	43,396	24,161	17,748	9,519	4,097	7				
2006	125,536	105,505	43,396	24,161	17,748	9,519	4,097	7			
2007	166,371	125,536	105,505	43,396	24,161	17,748	9,519	4,097	7		
2008	139,782	166,371	125,536	105,505	43,396	24,161	17,748	9,519	4,097	7	
2009	118,982	139,782	166,371	125,536	105,505	43,396	24,161	17,748	9,519	4,097	7
2010	104,338	118,982	139,782	166,371	125,536	105,505	43,396	24,161	17,748	9,519	4,097
2011	95,621	104,338	118,982	139,782	166,371	125,536	105,505	43,396	24,161	17,748	9,519
2012	146,021	95,621	104,338	118,982	139,782	166,371	125,536	105,505	43,396	24,161	17,748
2013		146,021	95,621	104,338	118,982	139,782	166,371	125,536	105,505	43,396	24,161
2014			146,021	95,621	104,338	118,982	139,782	166,371	125,536	105,505	43,396
2015				146,021	95,621	104,338	118,982	139,782	166,371	125,536	105,505
2016					146,021	95,621	104,338	118,982	139,782	166,371	125,536
2017						146,021	95,621	104,338	118,982	139,782	166,371
2018							146,021	95,621	104,338	118,982	139,782
2019								146,021	95,621	104,338	118,982
2020									146,021	95,621	104,338
2021										146,021	95,621
2022											146,021

Note: The values in this table are cumulative values from Table 4.6 in column 5. The values are based on the assumption that a vehicle is on the road for 11 years and is then scrapped.

Table 4.9 Attributable PHEVs on the road, by year and vehicle age

Year	*<1 year*	*1 year*	*2 years*	*3 years*	*4 years*	*5 years*	*6 years*	*7 years*	*8 years*	*9 years*	*10 years*
2011	2,725										
2012	12,967	2,725									
2013		12,967	2,725								
2014			12,967	2,725							
2015				12,967	2,725						
2016					12,967	2,725					
2017						12,967	2,725				
2018							12,967	2,725			
2019								12,967	2,725		
2020									12,967	2,725	
2021										12,967	2,725
2022											12,967

Note: The values in this table are cumulative values from Table 4.6 in column 6. The values are based on the assumption that a vehicle is on the road for 11 years and is then scrapped.

Table 4.10 Attributable EVs on the road, by year and vehicle age

Year	*<1 year*	*1 year*	*2 years*	*3 years*	*4 years*	*5 years*	*6 years*	*7 years*	*8 years*	*9 years*	*10 years*
2011	3,605										
2012	4,902	3,605									
2013		4,902	3,605								
2014			4,902	3,605							
2015				4,902	3,605						
2016					4,902	3,605					
2017						4,902	3,605				
2018							4,902	3,605			
2019								4,902	3,605		
2020									4,902	3,605	
2021										4,902	3,605
2022											4,902

Note: The values in this table are cumulative values from Table 4.6 in column 7. The values are based on the assumption that a vehicle is on the road for 11 years and is then scrapped.

Column 4 reports the sum of miles driven by the number of vehicles in column 1. The sum of miles driven for a given year in column 4 is obtained by multiplying the number of vehicles of each age (from Table 4.8) by the average annual number of miles driven by vehicles of that age (from Table 4.7) and summing across all vehicle ages for the given year. The sums of miles in columns 5 and 6 are obtained in the same way from the numbers of vehicles in Tables 4.9 and 4.10.[9]

Column 7 in Table 4.11 represents the fuel savings by year attributable to VTO based on the respective per-mile fuel savings in Table 4.12. For example, for 1999, 13.8 gallons of gasoline were saved for each thousand HEV miles attributable to VTO: 13.8 x 97 = 1,339 gallons of gasoline saved (rounded to 1 in thousands of gallons of gasoline saved). Column 8 in Table 4.11 is the value of the attributable fuel savings in column 7. This value is based on the average inflation-adjusted price of gasoline, by year, shown in Table 4.13. For example, for 1999, 1.339 thousand gallons of gasoline saved at an average price of $1.56 per gallon is $2.089 (rounded to $2 in thousands of 2012$).

The inflation-adjusted fuel cost savings in column 8 of Table 4.11 represents the annual economic benefits attributable to VTO's investments in NiMH and Li-ion battery technologies. A portion of these economic benefits are added to the environmental health benefits derived in Chapter 5. These environmental health benefit estimates are then compared with VTO's investments in Chapter 6 using the traditional evaluation metrics.

Table 4.11 Attributable fuel savings for U.S. HEVs, PHEVs, and EVs, 1999 through 2022

Year	*(1) Attributable HEVs on road (rounded)*	*(2) Attributable PHEVs on road (rounded)*	*(3) Attributable EVs on road (rounded)*	*(4) Attributable HEV miles driven (rounded, thousands)*	*(5) Attributable PHEV miles driven (rounded, thousands)*	*(6) Attributable EV miles driven (rounded, thousands)*	*(7) Attributable fuel savings (rounded, thousands of gallons)*	*(8) Attributable fuel cost savings (rounded, thousands 2012$)*
1999	7			97			1	2
2000	4,104			58,895			814	1,611
2001	13,623			195,901			2,707	5,027
2002	31,372			449,485			6,211	10,773
2003	55,533			788,913			10,901	21,419
2004	98,929			1,393,754			19,258	43,500
2005	204,434			2,877,430			39,759	106,175
2006	329,970			4,632,866			64,015	187,326
2007	496,341			6,918,298			95,594	295,234
2008	636,123			8,736,302			120,714	423,230
2009	755,105			10,178,950			140,648	356,448
2010	859,437			11,337,092			156,651	461,639
2011	950,961	2,725	3,605	12,304,344	39,099	51,735	172,295	627,142
2012	1,087,462	15,692	8,507	13,925,831	225,450	122,442	200,216	736,723
2013	1,069,714	15,692	8,507	13,342,510	225,385	121,129	192,115	706,913
2014	1,045,552	15,692	8,507	12,614,769	216,040	114,893	181,701	668,594
2015	1,002,156	15,692	8,507	11,573,278	201,471	108,484	166,852	613,955
2016	896,651	15,692	8,507	10,115,326	196,524	106,227	146,549	539,246
2017	771,115	15,692	8,507	8,445,693	193,353	103,869	123,349	453,881
2018	604,744	15,692	8,507	6,519,018	186,972	100,942	96,523	355,171
2019	464,962	15,692	8,507	4,933,139	182,880	97,979	74,446	273,935
2020	345,980	15,692	8,507	3,563,269	175,340	94,742	55,283	203,423
2021	241,642	15,692	8,507	2,411,963	166,038	84,273	38,889	143,097
2022	146,021	12,967	4,902	1,219,271	108,276	40,933	20,048	73,768
Total	12,111,937	172,610	93,581	148,536,395	2,116,827	1,147,649	2,125,539	7,308,232

Note: Column 1 was derived as the horizontal sum, by year, from Table 4.8. Columns 2 and 3 were derived in the same way for PHEVs and EVs.

Table 4.12 Gallons of gasoline saved per 1,000 attributable miles driven

Vehicle type	*Fuel economy*	*Gallons saved per 1,000 attributable miles driven*
ICV	23.5 miles per gallon	N/A
HEV	34.8 miles per gallon	13.8*
PHEV	40.8 miles per gallon equivalent	18.1
EV	82.3 miles per gallon equivalent	30.4

*Gallons saved per 1,000 miles are derived from fuel economy as follows: 13.8 = 1,000((1/23.5) – (1/34.8)). Likewise, 30.4 = 1,000((1/23.5) – (1/82.3)).

Sources: Average fuel economy for ICVs, HEVs, and EVs comes from Huo et al. (2009, p. 1798). Fuel economy for PHEV is a weighted average of HEV and EV fuel economy based on the assumption that a PHEV operates in charge-depleting mode (comparable to EV) for 25.6 percent of its miles driven and in charge-sustaining mode (comparable to HEV) for 74.4 percent of its miles driven. This assumption comes from the Greenhouse Gases, Regulated Emissions and Energy Use in Transportation (GREET) model, developed by Argonne National Laboratory. PHEV fuel economy was therefore calculated as a miles-weighted average of gallons per mile: 40.8 = 1/((0.256)(1/82.3) + (0.744)(1/34.8)).

Table 4.13 Inflation-adjusted price of gasoline per gallon, by year

Year	*(1) Average weekly U.S. price per gallon, all grades*	*(2) Price index (2012 = 100)*	*(3) Average price per gallon (2012$)*
1999	1.18	75.3	1.56
2000	1.52	76.9	1.98
2001	1.46	78.6	1.86
2002	1.39	79.9	1.73
2003	1.60	81.6	1.96
2004	1.89	83.9	2.26
2005	2.31	86.7	2.67
2006	2.62	89.5	2.93
2007	2.84	92.1	3.09
2008	3.30	94.1	3.51
2009	2.41	94.9	2.53
2010	2.84	96.2	2.95
2011	3.58	98.3	3.64
2012	3.68	100.0	3.68

Sources: Column 1 from www.eia.gov/oil_gas/petroleum/data_publications/wrgp/mogas_history.html; Column 2 from Table 1.1.

Note: The inflation-adjusted per gallon price of gasoline is assumed to stay at the 2012 value through 2022.

Appendix 4.1: Interview guide and responding organizations

The interview guide developed for this study is below. After each question we have annotated the guide with a number of paraphrased responses (in *italics*) that we received. These are included here to provide the reader with in-depth information about the expertise and representativeness of the interviewees. Because this study relies on expressed preferences and options rather than revealed preferences, we believe that emphasizing the background of those from whom opinions and estimates were obtained goes a long way to justifying the credibility of our quantitative findings.

***

The U.S. Department of Energy's EERE contracted with RTI International to conduct a benefit–cost study of its VTO's R&D investments in NiMH and Li-ion battery technologies. Toward that end, we deeply appreciate your taking the time to share your insights with us through your answers to the following questions.

Your responses will be kept confidential; only consensus opinions, without attribution, will be included in our final report to EERE and VTO.

1 Please provide a brief description of your current battery research, including how (if at all) this research has been affected (directly or indirectly) by VTO investment in NiMH or Li-ion battery technologies.

- *Our current battery research is focused on high-energy Li-ion cell technology development and electrochemical modeling, including performance and life prediction. The VTP investment in especially Li-ion battery technology has influence our research direction and progress to some extent by providing insight both on what works and what does not.*
- *We are currently investigating solid-state rechargeable Li-ion batteries targeting specific energies >325 Wh/kg. Our solid-state approach allows us to use multiple electrolytes in different parts of a cell, whereas liquid electrolytes would mix. Each electrolyte serves a different function (e.g., electrolyte A is stable at low voltages at the negative electrode while electrolyte B is stable at high voltages at the positive electrode). The approach allows us to push to higher voltages, and hence energies, than were possible with a single electrolyte which needed to be stable at both electrodes. This work has been entirely funded by the VTP.*
- *We are developing next-generation anode materials (i.e., new chemistries) for Li-ion that will first appear in consumer electronics and may within 10 years be scaled up for automotive application (as opposed to optimization of current cathodes for more immediate automotive application). Is the application of new Li-ion chemistries first to consumer electronics, almost as a proving ground prior to automotive application, typical? Yes, or if not typical then certainly not unusual. For example, Li iron phosphate was used first for power tools. Many Li-ion chemistries were used first in cell phones and laptops.*

- *Our research is focused on developing novel economical and scalable approaches to generating functional nanostructured architectures of alternative anode materials to replace graphite. Accordingly, the effort is directed at generating nanostructured heterostructures of silicon and silicon-carbon nanocomposites exhibiting specific capacity matching the theoretical capacity of silicon. We have developed novel cost-effective high-energy mechanical milling approaches to generate nanocomposites of Si and C. These first-generation systems exhibit a specific capacity of 700 mAh/g with capacity loss of 0.01%/cycle and a loading density of 2.2g/cc. These initial results paved the way for the next-generation nanocomposites comprising carbon nanotubes (CNTs) containing nanoparticles of silicon decorating the CNTs. These second-generation systems exhibit capacities of ~2500 mAh/g with first cycle irreversible loss of ~17% and capacity loss of 0.1%/cycle. Efforts are ongoing to develop novel interface control additives and surface control additives to generate nanocomposites with first cycle irreversible loss ~15% and coulombic efficiency ~99.99% and capacity loss of 0.01%/cycle.*
- *We are doing research on silicon anode technology to improve cycle time in Li-ion batteries. VTP $ accelerated our research. Our research is also leveraging attempts to use the silicon anode technology for the company so as to have a presence in the consumer electronics market.*
- *I just completed a multiyear, multimillion dollar project that was 50/50 with DOE. We developed a process to make a chemical that enables an increase in energy density of Li-ion cells. Holding size of the battery constant this process will result in a 30% increase in driving range. Or, holding driving range constant the battery could be smaller and lighter.*
- *Our current battery research is directly funded by VTP. One is High Energy Cathode Material for Lithium Ion Batteries while the other is High Energy Lithium Sulfur Battery as the Next Generation of Energy Storage Technology. With the support of VTP, we have systematically explored the material properties as well as the whole battery system. With VTP's encouragement, our battery project interacts frequently with universities, industries, and other national labs in terms of materials synthesis, testing, and characterization. The deep understanding gained from collaboration benefits each partner in the team and accelerates the market penetration of electric vehicles.*

2 Please provide a brief overview of any previous research related to NiMH or Li-ion battery technology. How (if at all) was this research affected (directly or indirectly) by VTO investment?

- *Our previous research focus was on NiMH battery technology for EV and later HEV applications. The focus was first on identifying the right technology option for our EV and then on improving the basic electrochemistry to improve life and further reduce cost for our HEV. The VTP investment, especially via the USABC, was critical to furthering our knowledge of NiMH battery technology, and assisted us in understanding which suppliers were best-in-class.*

- *Our previous work has focused on developing Li-ion batteries containing a single solid-state electrolyte. This design has allowed us to produce a robust Li metal battery with superior safety and cycle life compared to liquid electrolyte-based Li metal batteries. None of this work was funded by the VTP, but we did receive other DOE funding.*
- *It would not have been possible to do the work that I do without the Vehicle Technology Program and its investment in R&D. In the early years of my career (during the 90s), funding was very tight for battery research and VTP was really the only game in town for those of us interested in lithium metal or lithium ion batteries (there was just a little money from BES as well). Even so, the level of funding was low, and I was lucky to publish two papers a year, since I had to do everything myself (the proposal writing, the reports, the research, the equipment maintenance . . .), couldn't afford a postdoc! Now that the funding allows me to hire postdocs to help out, I can publish more and get more done, and more young people are being trained in the field as well.*
- *Our previous research prior to support by the DOE-BATT program was supported by the Office of Naval Research. This initial research funding laid the foundation for the development of novel anode materials to replace graphite. The DOE-BATT funding certainly helped accelerate the work. Funding from the BATT program provided the opportunity to develop multipronged approaches comprising novel synthetic routes coupled with development novel binders and processes to fabricate efficient electrodes so that the final system would display the desired electrochemical characteristics.*
- *I had industrial experience on NiMH battery manufacture which was not affected by VTP investment. I had program experience related to the Li-ion anode research funded by DARPA. The outcomes of VTP's investment on Li-ion indeed benefit our research significantly.*
- *I was heavily involved in Li-ion technology development from 1993 until the end of 2005. A significant part of the work was effectively funded by the DOE, through USABC and FreedomCar. Since 2006, I was working with thermal batteries, product used by the DOD. Showing Li-ion could be a high-power chemistry as well as high energy. Without DOE funding, we would have been 2 to 5 years behind.*
- *My previous research on Li-ion battery was also supported by VTP when I was still a graduate student in the university. With VTP investment, I got systematic training from fundamental sciences behind battery technology to battery manufacturing. After I got my Ph.D. degree, I further applied my knowledge gained from VTP to develop new energy storage technologies. I am also now supervising postdocs, post bachelor and interns in the lab and spreading the knowledge to other young scientists.*
- *My previous research was focused on the improvement of the thermal stability of Li-ion battery electrolytes for aerospace applications. This research benefited greatly from research conducted as part of the VTP investment. Developments in Li-ion technology supported by VTP greatly benefited aerospace Li-ion research.*

3 Shifting now to thinking about industry-wide battery research efforts, imagine a counterfactual scenario without VTO's support for NiMH and Li-ion battery technologies. Could you briefly describe how R&D performance, both in industry and in universities, might have been different?

- *In my experience, energy storage (battery) research would still have been pursued to some degree at national labs and universities without the VTP's support; however, the outcome would have been far less beneficial. Specifically, because the DOE worked closely with the U.S. automakers via the USABC, they were able to quickly develop a comprehensive set of meaningful battery goals and test and analyze methods against which progress could be objectively measured, and technology uniformly evaluated. Absent this government industry partnership, the required time to achieve the same level of progress, especially at the national labs and universities, may have increased by at least a factor of two.*
- *Without funding for battery technologies, commercialization would have lagged. The funding given to the large battery manufacturers (e.g., A123, Compact Power, etc.) led directly to introduction of commercial plug-in vehicles. Similarly in universities, without DOE funding of this nature, new material development which enables improved battery performance would essentially be nonexistent. It would have taken 5 to 8 years longer to reach the present state-of-the-art.*
- *Without VTP, a whole generation of battery scientists would have been lost, given the state of funding for battery research in the 90s. Although the funding level was still low, at least it was maintained, and some students were being trained in battery research at that time, mostly on funds from the Vehicle Technology Program (there was a little money from BES as well, but not much). When interest in batteries exploded in the mid-2000s, those students, who had by then taken jobs at universities and companies, were there to re-ignite the research, lead programs in batteries, etc., in the U.S. This residue of personnel is likely to make a difference when it comes to establishing a domestic Li-ion vehicular battery industry – the lack of interest that U.S. battery companies showed in Li-ion battery technologies means that we ceded the small consumer electronics market completely to Japan and Asia. This was not helped by the attitude of the U.S. car companies at the time (someone from GM actually said to me "only you greenies in California are interested in EVs." This was right before the Prius took off in popularity). If not for VTP, we probably could not compete in the vehicular battery field, either – as it is, it is touch and go, since Japan now has 20 years of experience manufacturing Li-ion batteries.*
- *I believe that NiMH and Li-ion research would have been pursued in the absence of VTP's support, but the research would have progressed significantly more slowly. Due to the massive resource requirements for successful battery material development, safety, performance testing, cost reduction, life modeling, and failure diagnostics, industry and university research without VTP support would not have been able to investigate the amount of materials that*

have been researched with VTP, nor would the screening studies have been as thorough. Lower-level studies could have been performed with private industry to accomplish similar types of results, but the outcomes would have been much lower quality and less accurate.

- *Without the advanced research component of VTP, you would be working on more applied chemistries, with the net result being that you would not have any of the advancement that we've seen in terms of energy densities or even the cost reduction that we've seen across the industry.*
- *Without DOE, our development would be multiple years behind. What took 5 years would have taken 8. We believe the effect may have been even larger for smaller companies who do not have the resources that we have.*
- *The high $/kWh battery and its low cycling performance have been the challenge for electric vehicle development and commercialization. Without strong DOE support the lithium-ion battery materials research would have been limited at a small-scale research level. For example, silicon has been well known as an active material hosting the highest Li capacity. However, its high volume expansion and bad cycling performance led to a confused conclusion – it cannot be commercially used. DOE funded projects promotes the R&D in silicon anode and result in dramatic progress in reduction of the silicon anode material production cost and cycling performance. We have scaled up the Si anode material production to ca. 400kg/month, which allows us to prepare the electrode with higher loading through formulation work and hence large pouch cells can be made in plant.*

4 The following three sets of graphs for battery life, energy density, and cost roughly describe how the state-of-the-art electric drive vehicle (EDV) battery technology has advanced over the past 15 years. Please indicate how you think the advancement of the state-of-the-art would have been different in the counterfactual scenario without VTO's support for NiMH and Li-ion battery technologies.

Please indicate how you think the graph would look in the counterfactual by clicking onthe dotted line and dragging/resizing as appropriate.

Note: At this level of abstraction, roughly describing the industry average state-of-the-art, we have shown one trend line for battery life, reflecting no significant difference between NiMH and Li-ion. If you wish to indicate different impacts on the different technologies, please copy the dotted line and indicate which technology each line represents.

- *In all of these graphs, the main point is that, without VTP investment, we would be significantly less advanced over the past 15 years in energy storage development.*
- *In my opinion, without the VTP funding the progress in Li-ion would have been very much lagging behind that of NiMH. Certainly, progress in developing the plug-in and the all-electric vehicle would have been very much thwarted, as shown in the plot above.*

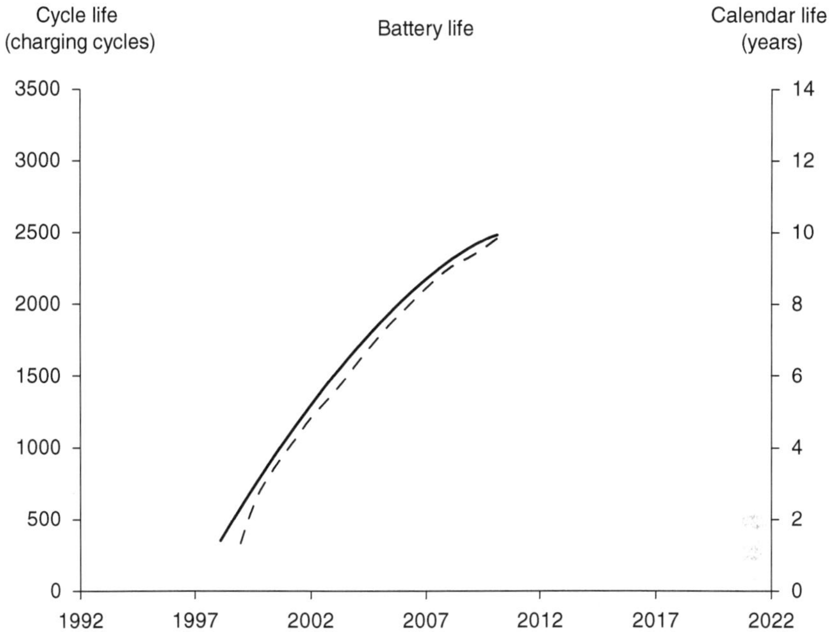

Figure A4.1 Charging cycles and calendar life (assuming full discharge)

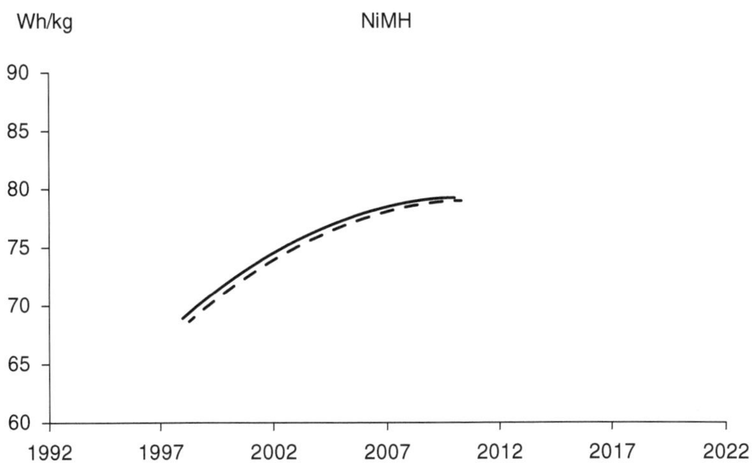

Figure A4.2a Energy density in NiMH batteries (Wh/kg)

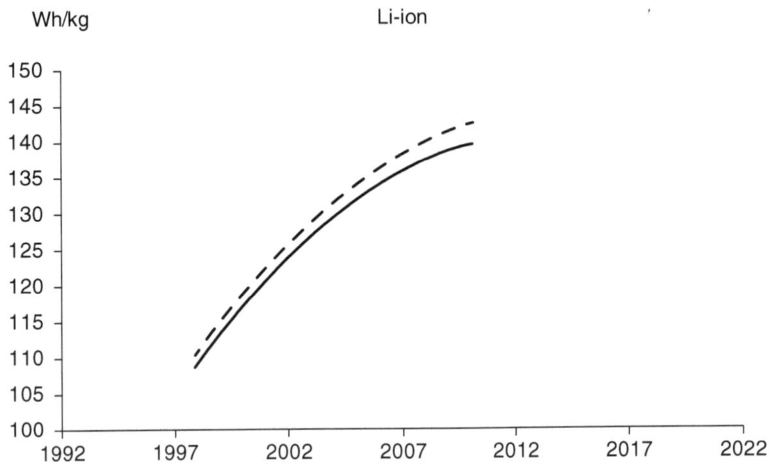

Figure A4.2b Energy density in Li-ion batteries (Wh/kg)

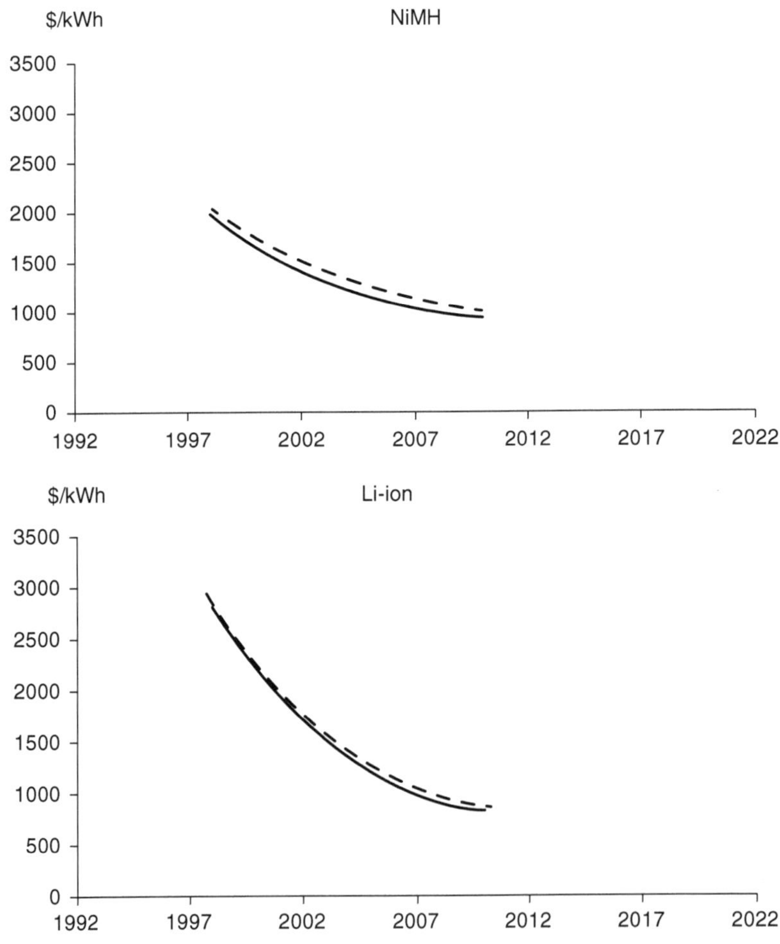

Figure A4.3 Cost in NiMH (top) and Li-ion (bottom) batteries ($/kWh)

5 The following graph presents a stylized depiction of the adoption of both NiMH and Li-ion technologies in the U.S. market through the purchase of EDVs. Please indicate (by clicking on the dotted line and dragging/resizing as appropriate and/or describing in words) how you think this adoption curve would have been different in the counterfactual scenario without VTO's support for NiMH and Li-ion battery technologies.

6 Li-ion technology made its way into the EDV market in a significant way relatively recently, making up roughly 4 percent of EDV sales in 2010 and 12 percent in 2011. In the counterfactual scenario without VTO's support for NiMH and Li-ion battery technologies, would the share of the EDV market represented by Li-ion have been different? If so, how?

- *It is possible that without the VTP's support for battery technology development, there might presently be no Li-ion technology in the EDV market. The influence on U.S. automakers is evident with all offering, or very soon to offer, EDV exclusively with Li-ion technology. By contrast, the Japanese automakers have been relatively slow to adopt Li-ion technology in vehicle applications, and still are not offering it in their HEV models.*
- *At this early adoption stage of Li-ion, I think the impact is ~0.25% decrease (3.75%) without VTP funding. As EDV market penetration increases, the technological and cost improvements from VTP support will play a more significant role.*
- *Without VTP support, I believe that the share of the EDV market would be significantly smaller since there would not be the infrastructure available to properly qualify battery technologies for electric vehicle applications with just industry and university support.*

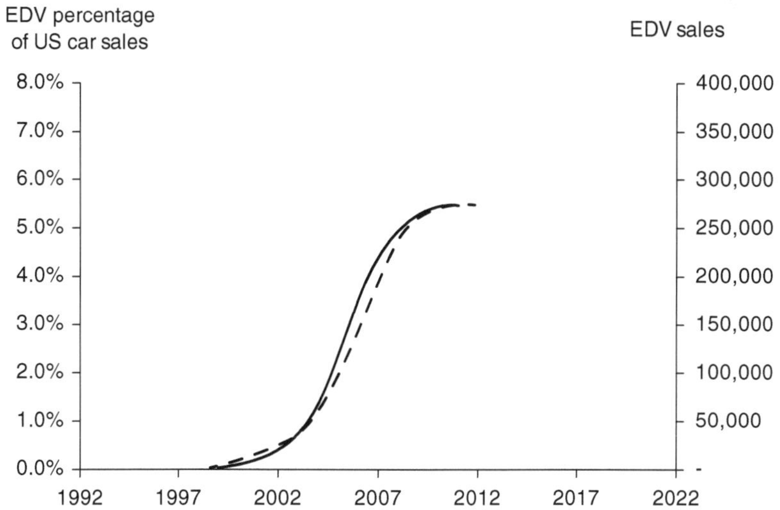

Figure A4.4 Battery technology adoption

- *As stated earlier, and as shown in the graph above, some of the advances made in the area of electrifying transportation would not have been possible, but for the support (both financial and technical) from the VTP. Some of the breakthrough chemistries for Li-ion were the result of VTP programs; the ARRA funds to set up production lines were a big determining factor for the industry to time the production of EDVs. Given the tough economy over the last few years, coupled with the higher cost of batteries in the event that the VTP programs were not in place, would have driven down the market penetration of EDVs back to pre-2000 levels.*
- *Without DOE there would be no Li-ion industry in the U.S. No Volt. No Li-ion EDVs by U.S. companies.*
- *Without DOE, there would be essentially no U.S. industry. Technology would still have been developed abroad in, e.g., Japan and Korea, and EDVs would still have made their way into the U.S. market, but it would have taken longer. How much longer? Guess: 10 years longer to where we are today.*

7 For clarification, are the differences that you have described between the actual and counterfactual scenarios due *entirely* to VTO's financial support of NiMH and Li-ion technologies?

- *The VTP's financial support has been instrumental, especially in encouraging the right type and appropriate review of energy storage technology research at national labs and universities. In addition, the DOE Office of EERE has strategically funded key R&D at battery developers via its cooperative agreements with the USABC. This combination of influences has been instrumental at moving the needle on both NiMH and Li-ion battery technology, and resulted in necessary domestic business investment and IP development. There are other important factors such as the acceptance of risk by the U.S. automakers when investing in new EDV and battery technology, but these too are enabled to some extent by the shared DOE investment.*
- *The effects of VTP supported projects are both direct and indirect. These projects generated tremendous indirect effects by their outreach efforts through workshops, seminars and demonstrations to educate the general public about these new technologies, which have significantly increased the market acceptance of the EDV powered by NiMH and Li-ion batteries.*
- *Yes, I believe that VTP's investment has enabled significant progress in the development of NiMH and Li-ion technologies for EDV applications by providing the infrastructure to investigate barriers to the successful commercialization of these technologies (cost, life, safety, abuse tolerance, etc.) and providing the necessary resources and expertise to significantly advance these technologies.*
- *Yes. I don't see any major other investment in the U.S. government over the last two decades. This is changing now and the future might look different. However, VTP is the program which will continue to bring technologies to a market-ready technology level and allows other more fundamental programs to become meaningful to the U.S. economy.*

- *Yes, without the level of funding provided by the VTP, the progress seen today would not have been possible at all. In fact, if the same extent of funding and the decision to focus on alternative energy storage had been in place since the 1990s, there would have been a lot more progress and there would have been a lot more EDVs providing better performance at a lower and more affordable price range. Despite this, I am confident that if the current level prevails, we will make a lot more progress in the years to come.*
- *Yes, for Li-ion. If it were not for VTP funding, all research and development of Lithium-ion batteries would have been performed in Asia only. No major American battery companies were developing or producing Lithium-ion batteries once Energizer closed its new Lithium-ion factory near Gainesville, FL over a decade ago. VTP kept America in the advanced battery game!*

Those organizations that participated in the survey are listed in Table A4.1.

Table A4.1 Participating organizations

Companies	*Universities*	*National laboratories*
A123 Systems (Navitas Systems)	Northwestern University	Argonne National Laboratory
Amprius	Pennsylvania State University	Brookhaven National Laboratory
Applied Materials, Inc.	SUNY Binghamton	Idaho National Laboratory
BASF Battery Materials, USA	University of Massachusetts Boston	Lawrence Berkeley National Laboratory
BASF Catalysts, LLC	University of Pittsburgh	National Renewable Energy Laboratory
Dow Kokam, LLC	University of Rhode Island	Oak Ridge National Laboratory
EnerDel, Inc.		Pacific Northwest National Laboratory
FMC Corporation		Southwest Research Institute
Ford Motor Company		U.S. Army Research Laboratory
General Motors		
H&T Waterbury		
K2 Energy Solutions, Inc.		
LG Chem Power, Inc.		
Maxwell Technologies, Inc.		
Miltec UV International		
Nanosys, Inc.		
Saft America, Inc.		
Seeo, Inc.		
Ultimate Membrane Technologies		

Note: Permission was given by those surveyed to list their affiliation.

Notes

1 In two instances, the interview was abbreviated because the scientists were deemed to be insufficiently qualified to offer an informed opinion about the economic impact of VTO's R&D investments.
2 For ease of response, a respondent could click on the curve and drag from the side or corner to stretch vertically and/or horizontally, and/or he/she could slide the curve without changing its shape.
3 See www.inemi.org/2013-roadmap.
4 We have no information as to why other potential interviewees declined to participate in this study.
5 This interview finding complements the information learned through the background interviews. See Box 4.1: VTO's investment increased the transition from lead acid battery technology to alternative battery technologies and the resulting adoption of EDVs on the road by 5 to 10 years.
6 None of those interviewed expressed any hesitance to respond to this data collection process, and the respondents were very cautious and conservative in the way that they discussed the impact of VTO's R&D investments.
7 To reiterate, emphasis on attribution to VTO's R&D investments during the discussion about Question 5 and again about Question 7 controls for several factors that are exogenous to VTO's R&D investments in battery technology. Those factors included, for example, global acceptance of EDVs, technical advancements by Japanese battery manufacturers, and other U.S. regulations (e.g., increased CAFE standards and tax credits for the purchase of EDVs).
8 Additional support for this assumption comes from U.S. EPA Tier 2 requirements "Current data indicate that passenger cars are driven approximately 120,000 miles in their first ten years of life. Trucks are driven further. Current regulatory useful lives are 10 years/100,000 miles for LDV/LLDTs and 11 years/120,000 miles for HLDTs" (U.S. EPA 2000, p. 6789).
9 For example, the figure of 97 (thousand attributable miles) at the top of column 4 of Table 4.11 is the product of 6.7575 attributable HEVs (rounded to 7 in Table 4.8) and 14,350 miles per year (from Table 4.7).

5 Measurement of environmental health and energy security benefits

This chapter discusses the estimates of the environmental emissions, environmental health, and energy security benefits from the on-road operation of EDVs.[1] Environmental health benefits are presented for three time periods: the retrospective time period of 1999 to 2012; the remaining vehicle lifespan, 2013 to 2022; and the full life cycle of these vehicles, 1999 to 2022.[2]

Approach to emissions estimation

Emissions from vehicles can be divided into two parts: well-to-pump (WtP) and pump-to-wheel (PtW); see Figure 5.1. This chapter focuses on the PtW, or vehicle operation, emissions reductions from EDVs. Estimates of impacts from well-to-wheel (WtW), or total emissions that also account for fuel production, feedstocks, distribution, and transportation, are included in the appendix to this chapter.

Under the operation of a gasoline internal combustion engine, vehicles emit a number of pollutants that contribute to climate change, air pollution, acid rain, visibility impairment, and surface water pollution (U.S. EPA, 2000). The ability of EDVs to operate on electric power for a portion of, or all of, the vehicle's operation reduces the amount of gasoline combusted and therefore the amount of tailpipe emissions. The impacts of EDVs on greenhouse gas (GHG) emissions, health impacts from reductions in air pollutants, and energy security benefits from reduced fuel consumption are based on the following assumptions:

- Each HEV, PHEV, or EV purchase offsets the purchase of a new ICV.
- All vehicle types use a fuel blend of 90.5 percent reformulated gasoline and 9.5 percent ethanol.[3]
- Emissions from brake and tire wear do not change by vehicle type.

The GREET model, developed by Argonne National Laboratory, was used to estimate emissions reductions. The GREET model estimates emissions factors for each vehicle type (ICV, HEV, PHEV, and EV) in 5-year intervals called vehicle model years. GREET provides emissions factors for GHGs and common air pollutants. Table 5.1 lists the specific pollutants relevant to this analysis.

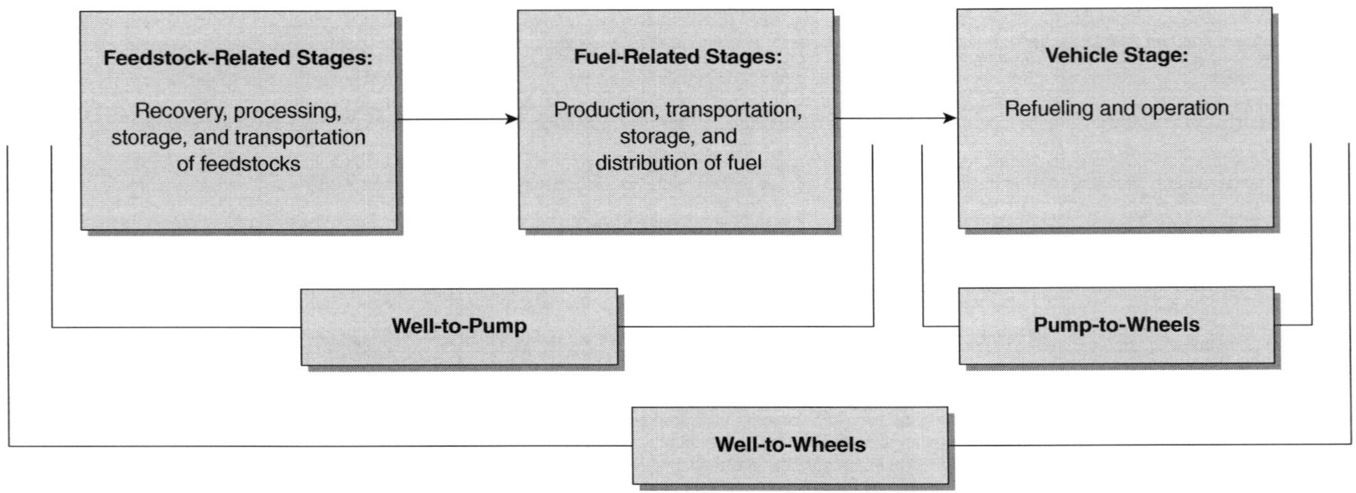

Figure 5.1 Well-to-wheels, well-to-pump, and pump-to-wheels analysis for fuel and vehicle systems

Source: Based on Wu et al. (2006).

Table 5.1 List of air pollutants

Name	*Abbreviation*	*Type*
Carbon dioxide	CO_2	Greenhouse gas
Methane	CH_4	Greenhouse gas
Nitrous oxide	N_2O	Greenhouse gas
Carbon monoxide	CO	Air pollutant that can oxidize into a greenhouse gas[a]
Volatile organic compounds	VOC	Air pollutant that can oxidize into a greenhouse gas[a]
Nitrogen oxides	NO_x	Air pollutant
Particulate matter smaller than or equal to 2.5 micrometers	$PM_{2.5}$	Air pollutant
Sulfur dioxide	SO_2	Air pollutant
Ammonia	NH_3	Air pollutant

Sources: U.S. EPA (2012a, e)

Note
[a]The carbon contained in CO and VOCs can oxidize into CO_2 (ANL 2012).

In 2000, the U.S. EPA instituted their Tier 2 requirements, which required reductions of nitrogen oxides (NO_x) and volatile organic compounds (VOCs) emissions as well as requiring lower sulfur content in gasoline for all vehicles. These requirements were phased in throughout the mid-2000s (U.S. EPA, 2000). To account for this adjustment in the analyses that follow, it was assumed that all vehicles purchased in 2005 and before possess the emissions factors from the GREET 2005 vehicle model year. All vehicles sold between 2006 and 2012 were assumed to possess the emissions factors of the GREET 2010 vehicle model year.[4] Figure 5.2 provides an overview of the methodology for this chapter.

Avoided greenhouse gas emissions

The combustion of gasoline during vehicle operations is a major contributor to GHG emissions. When an EDV operates on the electric drivetrain rather than using the gas engine, no tailpipe emissions occur, so it emits fewer GHGs into the atmosphere. These reductions will vary by vehicle type because an HEV operates using its gasoline engine more than a PHEV, while an EV will have zero tailpipe emissions because it does not use a gasoline engine. The three GHGs produced during fossil fuel combustion are carbon dioxide (CO_2), methane (CH_4), and nitrous oxide (N_2O).[5] Each has a differing impact depending on the gases' ability to absorb energy and how long the gas remains in the atmosphere; this is called the global warming potential (GWP). GWP is measured in equivalents of a metric ton of CO_2. Gasoline combustion emits far fewer quantities of CH_4 and N_2O than CO_2; however, CH_4 and N_2O are more effective at trapping heat in the atmosphere; thus, they both have a higher GWP than CO_2. CH_4 and N_2O contribute 20 and 300 times, respectively, more to global warming than CO_2 (U.S. EPA 2012a).

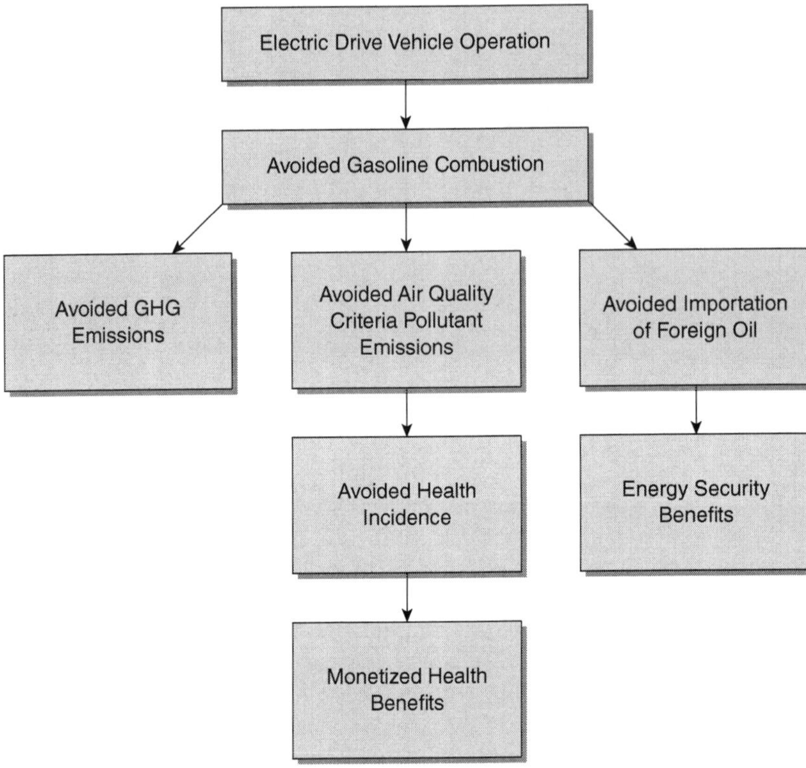

Figure 5.2 Approach for assessing environmental health benefits and energy security benefits from EDVs

The GREET model is used to estimate the rate of GHG emissions per mile driven by each vehicle type. The GREET model estimates a GHG emissions factor that captures the emissions of CO_2, CH_4, N_2O, and the carbon contained in VOCs and carbon monoxide (CO) that will fully oxidize into CO_2. This GHG estimate is measured in metric tons of equivalent CO_2 and accounts for the higher GWP of the non-CO_2 gases. Table 5.2 lists the emissions factors of GHGs by vehicle type and model year.

To calculate GHG emissions, the attributable miles from Table 4.11 were separated into the appropriate model year and vehicle type; see Table 5.3.

GHG emissions reductions were calculated by multiplying the attributable miles by the difference in emissions between the appropriate vehicle type and the conventional ICV to determine avoided GHG emissions. For example, in 1999 HEVs were driven 97,000 miles (rounded to the nearest thousand miles). This value was multiplied by 97.2 (the emissions factor of an ICV [337.2] subtracted

Table 5.2 Pump-to-wheels greenhouse gas emissions factors

	(1) ICV (g/mile)	(2) HEV (g/mile)	(3) PHEV (g/mile)	(4) EV (g/mile)
Models 1999–2005				
CO_2eq	337.6	240.4		
Models 2006–2012				
CO_2eq	318.7	227.0	163.1	0.0

Source: Based on ANL (2012).

Table 5.3 Attributable miles driven by vehicle type, 1999 through 2022

Year	(1) HEV (models 1999–2005; thousands)	(2) HEV (models 2006–2012; thousands)	(3) PHEV (thousands)	(4) EV (thousands)
1999	97			
2000	58,895			
2001	195,901			
2002	449,485			
2003	788,913			
2004	1,393,754			
2005	2,877,430			
2006	2,831,428	1,801,438		
2007	2,716,885	4,201,413		
2008	2,575,146	6,161,156		
2009	2,511,419	7,667,531		
2010	2,448,924	8,888,168		
2011	2,316,902	9,987,442	39,099	51,735
2012	2,140,903	11,784,928	225,450	122,442
2013	1,862,932	11,479,579	225,385	121,129
2014	1,528,189	11,086,580	216,040	114,893
2015	880,970	10,692,307	201,471	108,484
2016		10,115,326	196,524	106,227
2017		8,445,693	193,353	103,869
2018		6,519,018	186,972	100,942
2019		4,933,139	182,880	97,979
2020		3,563,269	175,340	94,742
2021		2,411,963	166,038	84,273
2022		1,219,271	108,276	40,933
Totals				
Retrospective (1999–2012)	23,306,083	50,492,075	264,549	174,176
Remaining lifespan (2013–2022)	4,272,091	70,466,145	1,852,278	973,473
Life-cycle (1999–2022)	27,578,175	120,958,220	2,116,827	1,147,649

Note: Numbers match those in Table 4.11. HEVs were broken out by model years before and after 2005.

from the emissions factor of an HEV [240.4]) and then converted to metric tons. Figure 5.3 shows avoided GHG emissions associated with EDVs sold from 1999 through 2012 for the years 1999 to 2022. Reductions peaked in 2012 at just over 1.3 million metric tons of equivalent CO_2 (CO_2eq).

During the retrospective time series, EDVs were responsible for avoiding 7.0 million metric tons of CO_2eq. When accounting for the full life cycle of these vehicles, the emission of 14.5 million metric tons of CO_2eq will be avoided. Table 5.4 shows the breakdown of GHG emissions reductions by vehicle type.

The majority of the avoided GHG emissions come from HEVs, which were responsible for over 98 percent of the GHG emissions during the retrospective time series and over 95 percent of the life-cycle time series. Although PHEVs and EVs have significantly lower PtW GHG emissions than HEVs and ICVs, these

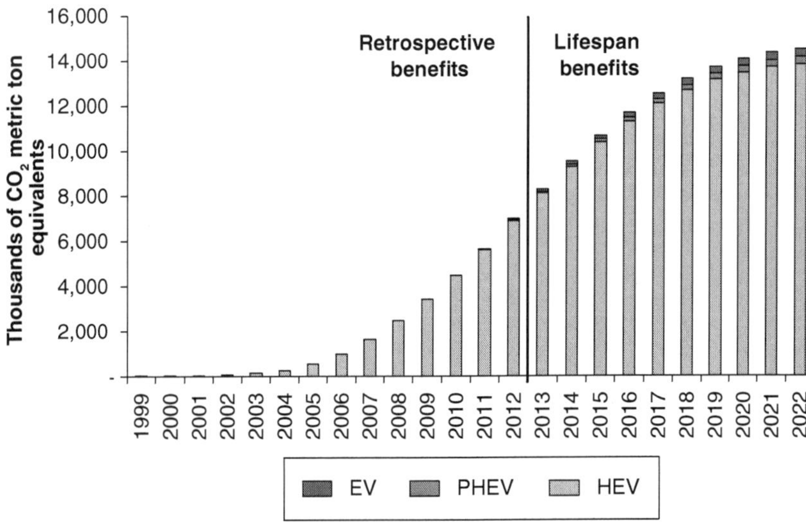

Figure 5.3 Cumulative pump-to-wheel avoided greenhouse gas emissions (thousands of metric tons of CO_2eq)

Table 5.4 Pump-to-wheel avoided greenhouse gas emissions, by vehicle type

Vehicle type	*1999–2012 (Thousands of metric tons of CO_2eq)*	*2013–2022 (Thousands of metric tons of CO_2eq)*	*1999–2022 (Thousands of metric tons of CO_2eq)*
HEV	6,893	6,873	13,766
PHEV	41	288	329
EV	55	310	366
Total	6,989	7,471	14,461

vehicles only account for a small percentage of the GHG reductions as a result of their comparatively lower sales volume through 2012.

The U.S. EPA's Greenhouse Gas Equivalency Calculator was used to convert these estimates of CO_2eq into everyday terms (U.S. EPA 2012b). The avoided 14.5 million metric tons of CO_2eq emissions from throughout the life-cycle time series is the equivalent of:

- the GHG emissions from the electricity used by 2,164,827 homes for 1 year;
- the carbon sequestered annually by 11,853,313 acres of U.S. forests; and
- the annual CO_2 emissions of 4.1 coal-fired power plants.

There are concerns that WtP emissions associated with the fuel source used for charging PHEV and EVs may cause a negative impact on the WtW, or total, GHG emissions. For example, CO_2 emissions would be higher if the electricity fuel source for a vehicle was a coal-fired power plant rather than a natural gas plant. Thus, CO_2 emissions in the WtP stage may be higher for a PHEV or EV than for an ICV and could increase GHG emissions. An assessment of the WtW stage avoided emissions for EDVs is in the appendix to this chapter.

WtW-avoided emissions of CO_2eq for PHEVs and EVs were calculated to be lower than the PtW stage but still positive. Ultimately, when accounting for the substantial avoided emissions of CO_2eq from HEVs, total WtW CO_2eq EDV emissions are higher than the PtW stage. It is important to note that GHG emissions from the WtP stage could become a real concern as sales of PHEVs and EVs increase, highlighting the importance of complementing clean energy technology on the roads with clean energy technology in the generation of electricity.

Current efforts are underway to account for the economic impact of climate change damages such as declined agricultural productivity or property damages from extreme weather events (U.S. EPA 2012c). The retrospective values of the Social Cost of Carbon (SCC), for nonregulatory analysis purposes, are still in their early stages of development. For this study, reductions in GHG emissions are not monetized. However, for the future, units of CO_2 and other GHG emissions may have monetary value based on social cost and/or CO_2 mitigation policies.

Avoided air pollutant emissions

The avoidance of fossil fuel combustion from using EDVs also reduces the emissions of several non-GHG pollutants. These pollutants do not trap heat in the atmosphere; however, they contribute to ozone and particulate matter (PM) pollution, which can negatively affect air quality and thus human health (U.S. EPA 2012e).[6] For example, when a vehicle emits VOC and NO_x, these pollutants interact in the lower atmosphere to form ambient ozone, which in high levels can exacerbate existing respiratory illness (such as asthma) and over the long term may cause lung damage (U.S. EPA, 2000). This study estimated the avoided emissions of VOCs, NO_x, PM smaller than or equal to 2.5 micrometers ($PM_{2.5}$), sulfur dioxide (SO_2), and ammonia (NH_3) from the reduced gasoline combustion of EDVs.

The analyses herein look at societal impacts; however, it is important to note that the majority of EDVs are purchased and operated in urban environments where air quality improvements are needed the most.[7] Because these impact estimates do not account for the higher marginal value of pollution abatement in urban areas where EDVs are most intensively used, they should be viewed as conservative lower bounds on actual impacts.

Emissions factors from the GREET model were used to determine avoided emissions in short tons. Similar to the GHG emissions calculated above, these emissions factors depend on vehicle type and model year; see Table 5.5. An estimate of NH_3 was provided from the U.S. EPA (2013) MOVES2010b model for a ICV, and reductions in NH_3 for EDVs were scaled based on the percentage reduction of VOCs (about 54 percent for HEVs) from the GREET model.

To determine the avoided emissions of air pollutants, the mileage estimates from Table 5.2 were multiplied by the difference in emissions between the appropriate vehicle type and the conventional ICV. Table 5.6 demonstrates the total avoided emissions of air pollutants from EDVs for the full life-cycle time series.

Table 5.5 Pump-to-wheels air quality criteria pollutant emissions factors

	(1) ICV (mg/mile)	*(2) HEV (mg/mile)*	*(3) PHEV (mg/mile)*	*(4) EV (mg/mile)*
Models 1999–2005				
VOC[a]	122.0	65.9		
NO_x	141.0	118.4		
$PM_{2.5}$	7.5	7.5		
VOC_{evap}	58.0	58.0		
SO_2[b]	5.6	4.0		
NH_3	17.0	9.2		
Models 2006–2012				
VOC	95.0	51.3	31.1	0.0
NO_x	69.0	58.0	35.1	0.0
$PM_{2.5}$	7.5	7.5	4.5	0.0
VOC_{evap}	57.0	57.0	34.6	0.0
SO_2	5.3	3.8	2.7	0.0
NH_3	17.0	9.2	5.6	0.0

Source: Based on ANL (2012) and U.S. EPA (2013)

Notes

[a]VOC emissions were derived from two sources: the emissions from vehicle exhaust and from the evaporation of gasoline in the fuel system (U.S. EPA 2000). These emissions are listed separately in this table and then are combined for an estimate of total avoided VOC emissions in subsequent tables.

[b]Emissions factors for SO_x were used as a proxy for SO_2 as for most combustion processes in GREET. SO_x is nearly 100 percent SO_2.

Table 5.6 Pump-to-wheel avoided air quality criteria pollutant emissions from EDVs, 1999 through 2022

Year	*(1)* VOC *(tons)*	*(2)* NO_x *(tons)*	*(3)* $PM_{2.5}$ *(tons)*	*(4)* SO_2 *(tons)*	*(5)* NH_3 *(tons)*
1999	0	0		0	0
2000	4	1		0	1
2001	12	5		0	2
2002	28	11		1	4
2003	49	20		1	7
2004	86	35		2	12
2005	178	72		5	25
2006	262	92		8	40
2007	370	119		12	60
2008	456	139		15	75
2009	525	156		17	88
2010	580	169		19	98
2011	637	185	1	21	108
2012	742	214	2	25	125
2013	710	204	2	24	120
2014	668	190	2	23	114
2015	607	168	2	21	104
2016	524	139	2	18	92
2017	443	118	1	15	77
2018	349	94	1	12	60
2019	271	74	1	9	47
2020	204	57	1	7	35
2021	146	42	1	5	24
2022	76	22	1	3	13
Total					
Retrospective (1999–2012)	3,928	1,217	2	128	643
Remaining lifespan (2013–2022)	3,998	1,107	14	137	686
Life-cycle (1999–2022)	7,926	2,324	16	265	1,329

Note: Totals may not add precisely because of rounding.

Similar to GHG emissions, HEVs account for over 90 percent of the VOC, NO_x, SO_2, and NH_3 avoided emissions. As with GHG emissions as discussed above, there are concerns that the WtP emissions associated with the fuel source used for charging PHEVs and EVs may cause there to be a negative impact on the WtW, or total, emissions of these air quality pollutants. For example, if the electricity fuel source is a coal-fired power plant, then SO_2 emissions in the WtP stage may be higher for a PHEV or EV than a conventional ICV and could increase overall emissions.

An assessment of the WtW stage avoided emissions for EDVs can be found in the appendix to this chapter. WtW emissions of SO_2 and $PM_{2.5}$ were found to increase from using PHEVs and EVs. This increase was offset by substantial declines in emissions of these pollutants from HEVs, resulting in a net reduction of emissions that was higher than the PtW stage. However, WtP emissions may become more of a concern as sales of PHEVs and EVs increase. Again, this highlights the need to complement clean energy technology on the roads with clean energy technology in the generation of electricity.

The majority of emissions reductions come in the form of VOC and NO_x reductions with 7,926 and 2,324 short tons avoided, respectively, throughout the life-cycle time series.

Reductions in $PM_{2.5}$ were small because only PHEVs and EVs were estimated to reduce $PM_{2.5}$ emissions in the PtW stage. Table 5.7 displays the total avoided emissions by vehicle type for the retrospective, lifespan, and life-cycle time series.

Environmental health benefits

The COBRA model, developed by EPA, was used to value the avoided health impacts from the reductions in air quality pollutants from EDVs. The COBRA model converts inputs of emissions changes in several air pollutants into changes in health effects and calculates the economic value of these changes (EPA 2012d; O'Connor et al. 2013). A discussion of COBRA is also included in the appendix to this chapter.

Table 5.7 Pump-to-wheels avoided air quality criteria pollutant emissions, by vehicle type

	(1) VOC (tons)	(2) NO_x (tons)	(3) $PM_{2.5}$ (tons)	(4) SO_2 (tons)	(5) NH_3 (tons)
Retrospective (1999–2012)					
HEV	3,874	1,194		126	636
PHEV	25	10	1	1	3
EV	29	13	1	1	3
Total	3,928	1,217	2	128	643
Remaining Lifespan (2013–2022)					
HEV	3,659	964		126	644
PHEV	176	69	6	5	23
EV	163	74	8	6	18
Total	3,998	1,107	14	137	686
Life-Cycle (1999–2022)					
HEV	7,533	2,158		252	1,280
PHEV	201	79	7	6	27
EV	192	87	9	7	22
Total	7,926	2,324	16	265	1,329

Note: Totals may not add precisely because of rounding.

COBRA's estimates of health impacts and their value are based on projected air quality, population, and income in 2017. For each year, the PtW avoided emissions from Table 5.6 were inputs into the COBRA model. The number of health incidents for each year was scaled to reflect changes in the national resident population based on U.S. Census estimates through 2012. The resident population from 2013 through 2022 was estimated by applying the percentage changes in population from the 2013 Annual Energy Outlook to the 2012 population estimate from the U.S. Census.

Several valuation endpoints in COBRA, including the value placed on premature mortality, are adjusted over time based on projected changes in income and elasticity estimates for how willingness to pay for health outcomes will change with time. The 2017 value for health impacts was adjusted to the years in our analysis based on COBRA's elasticity estimates and per capita real GDP changes based on BEA estimates through 2012. Income changes from 2013 through 2022 were estimated by applying the percentage changes in per capita real GDP from the 2013 Annual Energy Outlook to BEA's 2012 per capita real GDP estimate.

Table 5.8 shows the avoided incidence, or the reduced number of health incidents, and their associated cost estimate by COBRA for 2012. The COBRA model produces high and low estimates of total health benefits based on two

Table 5.8 Pump-to-wheels environmental health benefits associated with using EDVs, 2017

Health effect	*Avoided incidence*	*Benefit (thousands of 2012$)*
Adult mortality (low)	2.3	17,348
Adult mortality (high)	6.0	44,589
Infant mortality	0.0[a]	43
Nonfatal heart attacks (low)	0.3	34
Nonfatal heart attacks (high)	2.6	316
Resp. hospital admissions	0.9	26
Cardio-vascular (CDV) hospital admissions	0.8	34
Acute bronchitis	3.8	2
Upper respiratory symptoms	68.4	5
Lower respiratory symptoms	47.9	2
Asthma ER visits	1.6	1
Minor restricted activity days (MRAD)	2,114.6	147
Work loss days	357.3	56
Asthma exacerbations	71.9	8
Total health costs (2012$)		17,706 (Low) 45,229 (High)

Note

[a]The COBRA model rounded the incidence value to zero from a very small value; however, because of the high monetary value associated with prolonging life, even a very small value produces some cost.

different sets of assumptions for the impact of ambient $PM_{2.5}$ on adult mortality and nonfatal heart attacks. For adult mortality and nonfatal heart attacks, benefits or costs are expected to continue in future years, therefore COBRA calculates the present value for these endpoints using either a 3 percent or 7 percent discount rate. COBRA does not report undiscounted values.

In 2012 alone, the health benefits associated with using EDVs are estimated to be between \$17.7 and \$45.0 million using a 7 percent discount rate. The majority of these benefits come from avoided mortality, estimated to be between \$17.3 and \$44.6 million. The full time series of environmental health benefits is presented in Table 5.9. Note that the low and high environmental health benefits from the COBRA model are presented in Table 5.9. The mean of the low and high values is used in the economic analysis in Chapter 6.

PtW stage health benefits totaled between approximately \$88.2 and \$225.3 million for the retrospective time series and between \$192.0 and \$490.6 million for the life-cycle time series. An assessment of WtW health benefits is presented for reference in an appendix to this chapter.

Energy security benefits

The reduced consumption of gasoline from using EDVs not only reduces GHG and air quality pollutant emissions, but can also provide energy security benefits. Avoided fuel consumption can lead to a reduced demand for foreign oil importation, which may lessen the risk of oil disruptions and price shocks that could threaten the U.S. economy, environment, and national security.

To measure energy security benefits, we calculated the number of barrels of foreign oil avoided across the time series; see Table 5.10. The assumption in this time series is that each marginal barrel of oil is sourced from the same export nations as the total volume of imported crude oil.

From 1999 to 2012, approximately 27.2 million barrels of foreign oil were avoided from using EDVs. Accounting for the full life cycle of these vehicles, this study estimates a total of 47.8 million barrels of foreign oil will be avoided by 2022. These energy security benefits are not included in the economic analysis of this study because no consensus estimates of a so-called oil importation premium exist to monetize energy security benefits.

Table 5.9 Pump-to-wheels time series of environmental health benefits associated with using EDVs, 1999 through 2022

Year	*(1) EDV miles driven[a] (thousands)*	*(2) U.S. resident population[b] (thousands)*	*(3) Environmental health benefits (low; thousands $)*	*(4) Environmental health benefits (high; thousands $)*
1999	97	279,040	0	0
2000	58,895	282,162	68	174
2001	195,901	284,969	229	584
2002	449,485	287,625	531	1,357
2003	788,913	290,108	946	2,416
2004	1,393,754	292,805	1,703	4,351
2005	2,877,430	295,517	3,578	9,140
2006	4,632,866	298,380	5,548	14,172
2007	6,918,298	301,231	8,133	20,776
2008	8,736,302	304,094	10,169	25,975
2009	10,178,950	306,772	11,684	29,844
2010	11,337,092	309,326	13,140	33,563
2011	12,395,177	311,588	14,757	37,695
2012	14,273,723	313,914	17,706	45,229
2013	13,689,024	316,969	17,183	43,894
2014	12,945,702	320,051	16,492	42,129
2015	11,883,232	323,156	15,349	39,209
2016	10,418,077	326,283	13,634	34,830
2017	8,742,915	329,427	11,766	30,057
2018	6,806,933	332,586	9,475	24,206
2019	5,213,999	335,758	7,552	19,293
2020	3,833,351	338,942	5,831	14,896
2021	2,662,274	342,133	4,289	10,958
2022	1,368,481	345,328	2,271	5,802
Undiscounted totals[c]				
Retrospective (1999–2012)			88,192	225,277
Remaining lifespan (2013–2022)			103,842	265,274
Life-cycle (1999–2022)			192,034	490,552

Notes

[a]Sum of all columns in Table 5.3.

[b]Sources: 1999 through 2012 estimates from U.S. Census Bureau (2004, 2011, 2013). 2013 through 2022 based on percentage changes in population projected in the 2013 Annual Energy Outlook (U.S. Energy Information Administration 2013).

[c]The COBRA model applied a 7 percent discount rate to the stream of benefits associated with the emissions reduction in each year. Totals are undiscounted in the sense that no further discounting is applied here.

Table 5.10 Pump-to-wheels energy security benefits, 1999 through 2022

Year	*(1) Gallons of gasoline avoided*[a] *(thousands)*	*(2) Barrels of oil avoided*[b] *(thousands)*	*(3) Percentage of net foreign oil imports*[c]	*(4) Barrels of foreign oil avoided (thousands)*
1999	1	0	51%	0
2000	814	43	53%	23
2001	2,707	142	55%	79
2002	6,211	327	53%	174
2003	10,901	574	56%	322
2004	19,258	1,014	58%	592
2005	39,759	2,093	60%	1,262
2006	64,015	3,369	60%	2,017
2007	95,594	5,031	58%	2,924
2008	120,714	6,353	57%	3,612
2009	140,648	7,403	51%	3,804
2010	156,651	8,245	49%	4,053
2011	172,295	9,068	45%	4,051
2012	200,216	10,538	41%	4,314
2013	192,115	10,111	39%	3,916
2014	181,701	9,563	36%	3,436
2015	166,852	8,782	36%	3,157
2016	146,549	7,713	35%	2,684
2017	123,349	6,492	35%	2,249
2018	96,523	5,080	34%	1,750
2019	74,446	3,918	34%	1,338
2020	55,283	2,910	34%	997
2021	38,889	2,047	35%	708
2022	20,048	1,055	35%	372
Total				
Retrospective (1999–2012)	1,029,784	54,199		27,226
Remaining lifespan (2013–2022)	1,095,754	57,671		20,607
Life-cycle (1999–2022)	2,125,539	111,870		47,834

Notes

[a]Column 7 from Table 4.11.

[b]According to the U.S. EIA (2012b), on average, 19 gallons of motor gasoline can be yielded from 1 barrel of oil.

[c]Source: Based on U.S. EIA (2012a).

Appendix 5.1: Assessment of avoided emissions and overview of the COBRA model

WtW-avoided emissions

The WtW analysis accounts for emissions from feedstock and the distribution of fuel (WtP) in addition to those emissions from operating the vehicle (PtW). Although the emissions factors are different for the WtW analysis, avoided emissions were calculated in the same manner as the PtW analysis in Chapter 5. For energy security benefits, only PtW avoided barrels of foreign oil were calculated. See Table A5.1, Figure A5.1, and Tables A5.2 through A5.5.

Table A5.1 WtW GHG emissions factors (g/mile)

	GV	*HEV*	*PHEV*	*EV*
Models 1999–2005				
CO_2eq	453.1	322.9		
Models 2006–2012				
CO_2eq	425.2	303.1	333.8	283.6

Source: Based on ANL (2012).

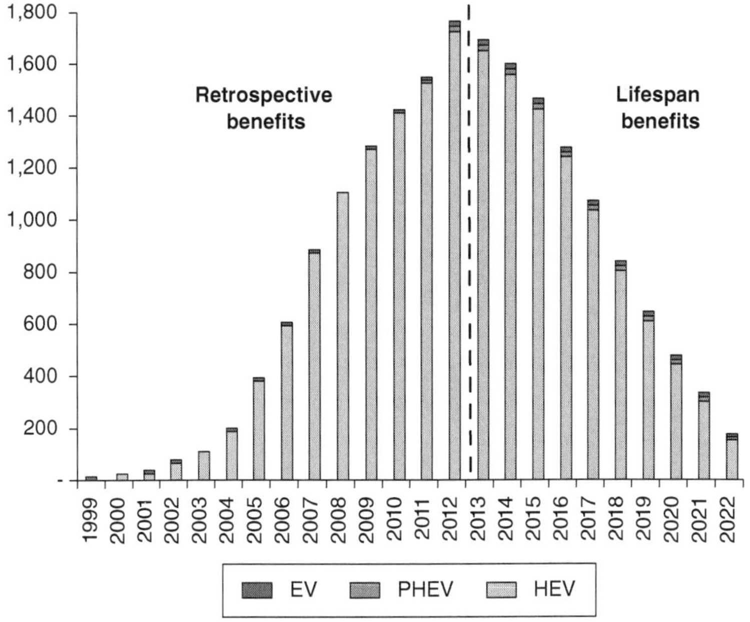

Figure A5.1 WtW-avoided GHG emissions (thousands of metric tons of CO_2eq)

Table A5.2 WtW-avoided GHG emissions by vehicle type (thousands of metric tons of CO_2eq)

Vehicle type	Retrospective	Lifespan	Life cycle
HEV	9,198	9,159	18,358
PHEV	24	169	193
EV	25	138	163
Total	9,247	9,466	18,714

Table A5.3 WtW air quality criteria pollutant emissions factors (mg/mile)

	(1) ICV	(2) HEV	(3) PHEV	(4) EV
Models 1999–2005				
VOC[a]	265.0	168.0	–	–
NO_x	389.5	295.9	–	–
$PM_{2.5}$	29.4	23.1	–	–
VOC_{evap}	58.0	58.0	–	–
SO_2[b]	147.2	105.2	–	–
NH_3[c]	36.9	23.4	–	–
Models 2006–2012				
VOC	228.9	147.0	112.9	23.4
NO_x	273.5	204.0	190.4	111.6
$PM_{2.5}$	26.9	21.3	62.6	119.2
VOC_{evap}	57.0	57.0	34.6	0.0
SO_2	126.8	90.6	355.8	719.8
NH_3	41.0	26.3	20.2	4.2

Sources: Based on ANL (2012) and U.S. EPA (2013).

Notes

[a]VOC emissions were derived from two sources: the emissions from vehicle exhaust and from the evaporation of gasoline in the fuel system (U.S. EPA 2000). These emissions are listed separately in this table and then are combined for an estimate of total avoided VOC emissions in subsequent tables.

[b]Emissions factors for SO_x were used as a proxy for SO_2 as for most combustion processes in GREET. SO_x is nearly 100 percent of SO_2.

[c]NH_3 WtW emissions for GV were scaled based on the percentage increase of VOC emissions from PtW to WtW. EDV reductions in NH_3 were scaled based on the percentage reduction of VOCs.

Overview of the COBRA model

The COBRA model provides estimates of health effect impacts and the economic value of these impacts resulting from emission changes. The COBRA model was developed by the U.S. EPA to be used as a screening tool that enables users to obtain a first-order approximation of benefits due to different air pollution mitigation policies.

Table A5.4 WtW-avoided air quality criteria pollutant emissions from EDVs (short tons), 1999 through 2022

Year	*(1)* VOC	*(2)* NO_x	*(3)* $PM_{2.5}$	*(4)* SO_2	*(5)* NH_3
1999	0	0	0	0	0
2000	6	6	0	3	1
2001	21	20	1	9	3
2002	48	46	3	21	7
2003	84	81	5	37	12
2004	149	144	10	65	21
2005	308	297	20	133	43
2006	465	430	30	203	71
2007	670	602	44	294	108
2008	832	737	55	365	138
2009	961	846	64	423	161
2010	1,065	933	71	468	180
2011	1,171	1,016	70	463	199
2012	1,364	1,166	65	433	233
2013	1,306	1,113	62	409	223
2014	1,231	1,047	58	384	212
2015	1,122	947	52	346	195
2016	975	811	43	285	172
2017	823	683	33	221	145
2018	647	534	22	147	114
2019	502	412	13	87	88
2020	376	306	5	36	65
2021	268	215	0	–1	46
2022	139	111	–1	–5	24
Total					
Retrospective analysis (1999–2012)	7,145	6,325	440	2,916	1,177
10-year lifespan analysis (2013–2022)	7,388	6,179	287	1,908	1,285
Life-cycle analysis (1999–2022)	14,533	12,504	728	4,824	2,462

At the core of the COBRA model is a source-receptor (S-R) matrix that translates changes in emissions to changes in PM concentrations. The changes in ambient PM concentrations are then linked to changes in mortality risk and changes in health incidents that lead to health care costs and/or lost workdays (O'Connor et al., 2013). Figure A5.2 provides an overview of the modeling steps.

Changes in emission → changes in ambient PM concentrations

The user provides changes (decreases) in emissions of pollutants ($PM_{2.5}$, SO_2, NO_x, VOCs, NH_3) and identifies the economic sector from which the emissions

Table A5.5 WtW-avoided air quality criteria pollutant emissions, by vehicle type (short tons)

	(1) VOC	(2) NO_x	(3) $PM_{2.5}$	(4) SO_2	(5) NH_3
Retrospective Analysis (1999–2012)					
HEV	7,054	6,269	469	3,097	1,164
PHEV	40	24	–10	–67	6
EV	50	31	–18	–114	7
Total	7,145	6,325	440	2,916	1,177
10-Year Lifespan Analysis (2013–2022)					
HEV	6,824	5,836	460	3,012	1,203
PHEV	283	170	–73	–468	42
EV	282	174	–99	–636	39
Total	7,388	6,179	287	1,908	1,285
Life-Cycle Analysis (1999–2022)					
HEV	13,878	12,105	928	6,109	2,367
PHEV	323	194	–83	–534	48
EV	332	205	–117	–750	47
Total	14,533	12,504	728	4,824	2,462

Note: Values may not add because of rounding.

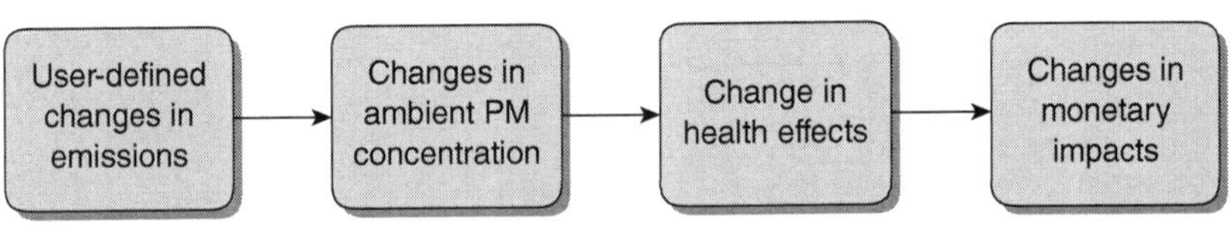

Figure A5.2 COBRA model overview

Source: Based on U.S. EPA (2012d).

are being reduced. These changes are in total short tons of pollutants by sector for the U.S. economy for the chosen analysis year. The economic sectors chosen determine the underlying spatial distribution of emissions and hence the characteristics of the human population that is affected.[8] For example, emissions reductions due to the use of EDVs are typically applied to light-duty gasoline vehicles. The S-R matrix consists of fixed transfer coefficients that reflect the relationship between annual average $PM_{2.5}$ concentration values at a single receptor in each county (a hypothetical monitor located at the county centroid) and the contribution by $PM_{2.5}$ species to this concentration from each emission source. This matrix provides quick but rough estimates of the impact of emission changes on ambient $PM_{2.5}$ levels as compared with the detailed estimates provided by more sophisticated air quality models (U.S. EPA, 2012d).

Changes in ambient PM concentrations → changes in health effects

The model then translates the changes in ambient PM concentration to changes in incidence of human health effects using a range of health impact functions and estimated baseline incidence rates for each health endpoint. The data used to estimate baseline incidence rates and the health impact functions used vary across the different health endpoints. To be consistent with prior U.S. EPA analyses, the health impact functions and the unit economic value used in COBRA are the same as the ones used for the Mercury and Air Toxics Standards (MATS) Final Rule (U.S. EPA, 2012d). The U.S. population estimates are based on projections from Woods and Poole (2011).

The model provides (in the form of a table or map) changes in the number of cases for each health effect between the baseline emissions scenario (included in the model) and the analysis scenario. The different health endpoints are included in Table A5.6.

Each health effect is described briefly below. For additional detail on the epidemiological studies, functional forms, and coefficients used in COBRA; see Appendix C of the COBRA user's manual (U.S. EPA 2012d).

Table A5.6 Health endpoints included in COBRA

Health effect	*Description*
Mortality	Number of deaths (adult or infant)
Acute bronchitis	Cases of acute bronchitis
Nonfatal heart attacks	Number of nonfatal heart attacks
Respiratory hospital admissions	Number of cardiopulmonary-, asthma-, or pneumonia-related hospitalizations
CDV-related hospital admissions	Number of cardiovascular-related hospitalizations
Upper respiratory symptoms	Episodes of upper respiratory symptoms (runny or stuffy nose; wet cough; and burning, aching, or red eyes)
Lower respiratory symptoms	Episodes of lower respiratory symptoms: cough, chest pain, phlegm, or wheeze
Asthma emergency room visits	Number of asthma-related emergency room visits
MRAD	Number of minor restricted activity days (days on which activity is reduced but not severely restricted; missing work or being confined to bed is too severe to be MRAD)
Work loss days	Number of work days lost due to illness
Asthma exacerbations	Number of episodes with cough, shortness of breath, wheeze, and upper respiratory symptoms in asthmatic children

Mortality researchers have linked both short-term and long-term exposures to ambient levels of air pollution to increased risk of premature mortality. COBRA uses mortality risk estimates from an epidemiological study of the American Cancer Society cohort conducted by Krewski et al. (2009) and by a Six-City cohort by Laden et al. (2006). These two studies provide a high and low estimate of mortality associated with changes in ambient $PM_{2.5}$. COBRA includes different mortality risk estimates for both adults and infants. Infant mortality is based on Woodruff et al. (1997). Because of the high monetary value associated with prolonging life, mortality risk reduction is consistently the largest health endpoint valued in the study.

Nonfatal heart attacks were linked by Peters et al. (2001) to PM exposure. Nonfatal heart attacks were modeled separately from hospital admissions because of their lasting impact on long-term health care costs and earning. COBRA provides a high and low estimate of incidence for nonfatal heart attacks based on differing literature.

Hospital admissions include two major categories: respiratory (such as pneumonia and asthma) and cardiovascular (such as heart failure, ischemic heart disease). Using detailed hospital admission and discharge records, Sheppard et al. (1999) investigated asthma hospital emissions associated with PM, CO, and ozone, and Moolgavkar (2000, 2003) found a relationship between hospital admissions and PM. COBRA includes separate risk factors for hospital admissions for people aged 18 to 64 and aged 65 and older.

Acute bronchitis, defined as coughing, chest discomfort, slight fever, and extreme tiredness lasting for a number of days, was found by Dockery et al. (1996) to be related to sulfates, particulate acidity, and, to a lesser extent, PM. COBRA estimates the episodes of acute bronchitis in children aged 8 to 12 from pollution using the findings from Dockery et al.

Upper respiratory symptoms include episodes of upper respiratory symptoms (runny or stuffy nose; wet cough; and burning, aching or red eyes). Pope et al. (2002) found a relationship between PM and the incidence of a range of minor symptoms, including runny or stuffy nose; wet cough, and burning; aching or red eyes.

Lower respiratory symptoms in COBRA are based on Schwarz and Neas (2000) and focus primarily on children's exposure to pollution. Children were selected for the study based on indoor exposure to PM and other pollutants resulting from parental smoking and gas stoves. Episodes of lower respiratory symptoms are coughing, chest pain, phlegm, or wheezing.

Asthma related emergency room visits are primarily associated with children under the age of 18. Sheppard et al. (2003) found significant associations between asthma ER visits and PM and CO. To avoid double counting, hospitalization costs (discussed above) do not include the cost of admission to the emergency room.

Minor restricted activity days (MRAD) in COBRA were based on research by Ostro and Rothschild (1989). MRADs include days on which activity is reduced but not severely restricted (e.g., missing work or being confined to bed is too severe to be an MRAD). They estimated the incidence of MRADs for a national sample of the adult working population, aged 18 to 65, in metropolitan areas. Because their study is based on a convenience sample of nonelderly individuals, the impacts may be underestimated because the elderly are likely to be more susceptible to PM-related MRADs.

Work loss days were estimated by Ostro (1987) to be related to PM levels. Based on an annual national survey of people aged 18 to 65, Ostro found that 2-week average PM levels were significantly linked to work loss days. However, the findings showed some variability across years.

Asthma exacerbations estimates were pooled from Ostro et al. (2001) and Mar et al. (2004) to calculate impacts of changes in air quality on asthmatic children. Cough, wheeze, and shortness of breath are all considered to be exacerbations.

Changes in health effects → changes in monetary impacts

COBRA translates the health effects into changes in monetary impacts using estimated unit values of each health endpoint. The per-unit monetary values are described in Appendix F of the COBRA user's manual (U.S. EPA 2012d). Estimation of the monetary unit values varies by the type of health effect. For example, reductions in the risk of premature mortality are monetized using value of statistical life estimates. Other endpoints such as hospital admissions use cost of illness units that include the hospital costs and lost wages of the individual but do not capture the social (personal) value of pain and suffering. COBRA allows users to choose between a discount rate of 3 percent or 7 percent to calculate the present value of health effects that may occur beyond the year 2017.

Limitations

It should be noted that COBRA does not incorporate effects of many pollutants, such as carbon emissions or mercury. This has two potential implications. First, other pollutants may cause or exacerbate health endpoints that are not included in COBRA. This would imply that reducing incidences of such health points are not captured. Second, pollutants other than those included in COBRA may also cause a higher number of incidences of the health effects that are part of the model. This is also not captured in this analysis. Thus, the economic value of health effects obtained from COBRA may be interpreted as a conservative estimate of the health benefits from reducing emissions.

Notes

1 This chapter draws from RTI International (2013), which was written by Sara Casey and Ross Loomis, previously at RTI International.

2 The full life of an EDV is assumed to be 11 years (Table 4.7). EDV sales were categorized by calendar year (January to December).

3 According to U.S. DOE (2013), over 95 percent of U.S. gasoline contains up to 10 percent ethanol, or E10.

4 Because PHEV and EVs were not introduced to the market until 2011, only the GREET 2010 model year estimates were used for these vehicle types. This assumption will account for the greater avoided reductions associated with early HEV models before the U.S. EPA's regulations took effect on all vehicles.
5 Hydrofluorocarbons, perfluorocarbons, and sulfur hexafluoride are also considered GHGs. However, these are not directly associated with fossil fuel combustion and therefore are not included in this analysis.
6 Ozone and PM also affect the environment through reduced visibility, reduction in agricultural and forestry yields, increased acid deposition, and increased eutrophication and nitrification of surface waters (U.S. EPA 2000). For the purposes of this study, only the impacts on human health were monetized.
7 Huo et al. (2009) found that HEVs and EVs can reduce emissions compared with ICVs of VOC, NO_x, and $PM_{2.5}$ at a higher percentage in urban areas than when applied to a national model.
8 The COBRA model has a variety of spatial capabilities. However, for this study there was limited information on the specific location of pollution reductions. Thus, a national analysis was conducted where the national distribution of emissions was used to determine the emission location as input to the S-R matrix.

6 Comparison of benefits and costs of VTO's R&D investments

This chapter compares VTO's R&D investments from Chapters 1 and 2 with the economic and environmental health benefits, attributable to those investments, from Chapters 4 and 5. The benefits are attributable to VTO investments by way of their association with the impact of those investments (as established through interviews, by methods described in Chapter 4) on the diffusion of EDVs (presented in Chapter 3). To make an appropriate comparison, the streams of costs and benefits are discounted to a common date: January 1, 1992, the beginning of the stream of investment costs.

We present three impact metrics: net present value (NPV), benefit-to-cost ratio (BCR), and internal rate of return (IRR). We address the robustness of our estimates of these metrics in three ways: by applying both 7 percent and 3 percent discount rates to present-value calculations (for NPV and BCR), by providing estimates using the upper and lower bounds of confidence intervals based on the variability of respondents' opinions, and by considering a truncated stream of benefits that stops on December 31, 2012, and thus does not include any benefits associated with the remaining useful lives of attributable EDVs beyond that date.

We find that benefits exceed costs (with BCR = 1.73) even under the most conservative scenario: using a 7 percent discount rate, confidence-interval lower bound, and considering only retrospective benefits (those accruing prior to December 31, 2012). A more optimistic and still defensible assessment can be based on the central tendency of respondents' opinions and consider the full useful lives of attributable EDVs. In this scenario, the IRR is 17.7 percent, and a 7 percent discount rate yields a BCR of 3.63.

The chapter develops these estimates step-by-step, based on the information presented in earlier chapters, and presenting all arithmetic formulas used. The chapter concludes with a discussion of why even the more-optimistic defensible assessment (IRR = 17.7 percent and 7 percent BCR = 3.63) may be considered rather conservative.

VTO's R&D investments

Column 1 in Table 6.1 is reproduced from column 4 in Table 1.1. VTO R&D investments, in 2012 dollars (2012$), are in column 1 over the years 1992

Table 6.1 VTO R&D investments in energy storage technology, 1992 through 2012

Year	*(1)* *VTO R&D investments (thousands of 2012$, rounded)*	*(2)* *VTO R&D investments discounted at 7 percent to 1/1/1992 (thousands of 2012$)*	*(3)* *VTO R&D investments discounted at 3 percent to 1/1/1992 (thousands of 2012$)*
1992	39,783	39,783	39,783
1993	45,557	42,577	44,230
1994	51,699	45,156	48,731
1995	39,200	31,999	35,874
1996	37,145	28,338	33,003
1997	34,763	24,786	29,987
1998	37,397	24,919	31,319
1999	30,272	18,852	24,614
2000	30,474	17,736	24,056
2001	35,234	19,165	27,004
2002	33,375	16,966	24,834
2003	30,051	14,277	21,709
2004	26,987	11,983	18,928
2005	29,832	12,379	20,314
2006	31,444	12,195	20,788
2007	48,238	17,484	30,962
2008	52,128	17,658	32,484
2009	77,347	24,486	46,796
2010	82,035	24,271	48,187
2011	84,778	23,442	48,348
2012	93,034	24,042	51,511
Totals	970,773	492,491	703,463

Note: Investment costs are assumed to have occurred at the beginning of each time period.

through 2012. These costs represent VTO's total R&D investments in all areas of energy storage, including NiMH and Li-ion battery technologies. We emphasize that these investments overstate the relevant costs associated with the economic and energy benefits and environmental health benefits measured and monetized in Chapters 4 and 5, respectively, because investments are included that did not specifically target NiMH and Li-ion technology. Also, the significant lag between R&D investment and returns is ignored when investments through 2012 are counted against the benefits associated with attributable numbers of vehicles sold through 2012, a point revisited at the end of this chapter.

Columns 2 and 3 in Table 6.1 show the present value of VTO's R&D investments (2012$), discounted to January 1, 1992, by 7 percent and 3 percent discount rates, respectively.[1] Formally, the discounted totals provided at the bottom of columns 2 and 3 are derived from the undiscounted costs in column 1 by the following formula, where r is the discount rate and C_y is the undiscounted VTO investment in year y:

$$\sum_{y=1992}^{2012} C_y (1+r)^{1992-y}$$

Economic and environmental health benefits attributable to VTO's R&D investments

Column 1 in Table 6.2 is reproduced from column 8 in Table 4.11. Columns 2 and 3 show these attributable fuel savings discounted to January 1, 1992, using 7 percent and 3 percent discount rates. Formally, the discounted totals provided at the bottom of columns 2 and 3 are derived from the undiscounted values in column 1 by the following formula, where r is the discount rate, and B_y is the undiscounted value of attributable fuel savings in year y, and y_stop is either 2012 or 2022 ($y+1$ is used in the exponent because benefits are assumed to accrue at the end of each year, whereas costs are incurred at the beginning of each year):

$$\sum_{y=1999}^{y\_stop} B_y (1+r)^{1992-(y+1)}$$

Column 1 in Table 6.3 presents the average of the environmental health benefits in columns 3 and 4 in Table 5.9 (recall that these values were generated by COBRA, applying a 7 percent discount rate). Column 2 in Table 6.3 presents values analogous to those in column 1 but generated by applying a 3 percent discount rate in COBRA; applying the lower discount rate to the streams of benefits in COBRA results in values that are 12 percent higher for each year. Column 3 in Table 6.3 discounts the values in column 1 to January 1, 1992, by applying a 7 percent discount rate; column 4 discounts the values in column 2 to January 1, 1992, by applying a 3 percent discount rate. Discounted totals are derived from undiscounted yearly values using the same formula provided above for economic benefits.

Table 6.4 presents the combined economic and environmental health benefits from Tables 6.2 and 6.3, respectively. Column 1 in Table 6.4 adds, for each year and total, column 1 of Table 6.2 to column 1 of Table 6.3; column 2 in Table 6.4 adds column 1 of Table 6.2 to column 2 of Table 6.3.

Economic evaluation metrics

Three economic evaluation metrics were calculated using the data on attributable total benefits in Table 6.4: NPV, BCR, and IRR. Each of these metrics was discussed at a conceptual level in Chapter 1.

Table 6.2 Attributable economic and energy benefits, 1999 through 2022

Year	*(1) Attributable fuel savings (thousands of 2012$)*	*(2) Attributable fuel savings discounted at 7 percent to 1/1/1992 (thousands of 2012$)*	*(3) Attributable fuel savings discounted at 3 percent to 1/1/1992 (thousands of 2012$)*
1999	2	1	2
2000	1,611	876	1,235
2001	5,027	2,555	3,741
2002	10,773	5,118	7,783
2003	21,419	9,510	15,023
2004	43,500	18,051	29,621
2005	106,175	41,176	70,194
2006	187,326	67,896	120,237
2007	295,234	100,006	183,980
2008	423,230	133,984	256,061
2009	356,448	105,460	209,376
2010	461,639	127,647	263,266
2011	627,142	162,065	347,233
2012	736,723	177,928	396,025
2013	706,913	159,560	368,933
2014	668,594	141,038	338,771
2015	613,955	121,039	302,025
2016	539,246	99,356	257,547
2017	453,881	78,156	210,462
2018	355,171	57,158	159,894
2019	273,935	41,200	119,731
2020	203,423	28,594	86,322
2021	143,097	18,798	58,954
2022	73,768	9,057	29,506
Undiscounted totals			
Retrospective (1999–2012)	3,276,249		
Remaining lifespan (2013–2022)	4,031,983		
Life-cycle (1999–2022)	7,308,232		
Discounted totals			
Retrospective (1999–2012)		952,275	1,903,777
Remaining lifespan (2013–2022)		753,955	1,932,145
Life-cycle (1999–2022)		1,706,230	3,835,922

Note: For purposes of discounting, economic benefits are assumed to be realized at the end of each time period.

Table 6.3 Attributable mean environmental health benefits, 1999 through 2022

Year	*(1) Attributable environmental health benefits (thousands of 2012$, applying a 7 percent discount rate in COBRA)*	*(2) Attributable environmental health benefits (thousands of 2012$, applying a 3 percent discount rate in COBRA)*	*(3) Attributable environmental health benefits discounted at 7 percent to 1/1/1992 (thousands of 2012$)*	*(4) Attributable environmental health benefits discounted at 3 percent to 1/1/1992 (thousands of 2012$)*
1999	0	0	0	0
2000	121	136	66	104
2001	406	455	207	339
2002	944	1,058	448	764
2003	1,681	1,885	746	1,322
2004	3,027	3,394	1,256	2,311
2005	6,359	7,129	2,466	4,713
2006	9,860	11,054	3,574	7,095
2007	14,455	16,205	4,896	10,099
2008	18,072	20,260	5,721	12,258
2009	20,764	23,278	6,143	13,674
2010	23,351	26,179	6,457	14,930
2011	26,226	29,402	6,777	16,279
2012	31,468	35,279	7,600	18,964
2013	30,539	34,237	6,893	17,868
2014	29,311	32,861	6,183	16,650
2015	27,279	30,583	5,378	15,045
2016	24,232	27,168	4,465	12,975
2017	20,912	23,445	3,601	10,871
2018	16,840	18,881	2,710	8,500
2019	13,422	15,049	2,019	6,577
2020	10,363	11,619	1,457	4,931
2021	7,624	8,548	1,002	3,522
2022	4,037	4,526	496	1,810
Undiscounted totals				
	156,734	175,715	Retrospective (1999–2012)	
	184,558	206,916	Remaining lifespan (2013–2022)	
	341,293	382,631	Life-cycle (1999–2022)	
Discounted totals				
Retrospective (1999–2012)			46,358	102,852
Remaining lifespan (2013–2022)			34,202	98,750
Life-cycle (1999–2022)			80,560	201,602

Note: Environmental health benefits are assumed to be realized at the end of each time period.

Table 6.4 Attributable total economic and energy and environmental health benefits, 1999 through 2022

Year	(1) Attributable total benefits (thousands of 2012$, applying a 7 percent discount rate in COBRA)	(2) Attributable total benefits (thousands of 2012$, applying a 3 percent discount rate in COBRA)	(3) Attributable total benefits discounted at 7 percent to 1/1/1992 (thousands of 2012$)	(4) Attributable total benefits discounted at 3 percent to 1/1/1992 (thousands of 2012$)
1999	2	2	1	2
2000	1,732	1,747	942	1,339
2001	5,433	5,482	2,762	4,079
2002	11,717	11,831	5,567	8,547
2003	23,100	23,303	10,257	16,344
2004	46,527	46,894	19,307	31,932
2005	112,535	113,305	43,643	74,908
2006	197,186	198,380	71,469	127,333
2007	309,689	311,439	104,902	194,079
2008	441,302	443,491	139,705	268,319
2009	377,212	379,726	111,603	223,049
2010	484,990	487,818	134,104	278,196
2011	653,367	656,543	168,843	363,512
2012	768,191	772,002	185,528	414,989
2013	737,452	741,151	166,453	386,801
2014	697,904	701,454	147,221	355,421
2015	641,234	644,538	126,417	317,070
2016	563,478	566,414	103,820	270,522
2017	474,793	477,327	81,757	221,334
2018	372,011	374,051	59,868	168,394
2019	287,358	288,984	43,219	126,308
2020	213,786	215,042	30,050	91,252
2021	150,720	151,644	19,800	62,475
2022	77,805	78,294	9,552	31,317
Undiscounted totals				
	3,432,983	3,451,964	Retrospective (1999–2012)	
	4,216,541	4,238,899	Remaining lifespan (2013–2022)	
	7,649,525	7,690,863	Life-cycle (1999–2022)	
Discounted totals				
Retrospective (1999–2012)			998,633	2,006,628
Remaining lifespan (2013–2022)			788,158	2,030,895
Life-cycle (1999–2022)			1,786,791	4,037,523

Notes

Economic and environmental health benefits are assumed to be realized at the end of each time period.

Column 1 is the sum of column 1 in Table 6.2 and column 1 in Table 6.3.

Column 2 is the sum of column 1 in Table 6.2 and column 2 in Table 6.3.

Column 3 is the sum of column 2 in Table 6.2 and column 3 in Table 6.3.

Column 4 is the sum of column 3 in Table 6.2 and column 4 in Table 6.3.

Mathematically:

$$NPV = \left\{ \sum_{y=1999}^{2022} B_y (1+r)^{1992-(y+1)} \right\} - \left\{ \sum_{y=1992}^{2012} C_y (1+r)^{1992-y} \right\} \text{ and}$$

$$BCR = \left\{ \sum_{y=1999}^{2022} B_y (1+r)^{1992-(y+1)} \right\} \Big/ \left\{ \sum_{y=1992}^{2012} C_y (1+r)^{1992-y} \right\}$$

where, B_y is the value in 2012 dollars of the sum of economic benefits plus environmental health benefits realized in year y, C_y is the value in 2012 dollars of R&D investments made in year y, and r is the discount rate (analogous to a real interest rate used to reflect the opportunity cost of investment capital). Note that benefits are discounted an extra year, following the EERE guidelines that benefits for a given year are assumed to have been realized at year-end, while costs are assumed to have been incurred at year start (Ruegg and Jordan, 2011). These formulas convert benefits and costs to January 1, 1992, present value.

The IRR is the discount rate r^* such that NPV = 0 and BCR = 1:

$$\left\{ \sum_{y=1999}^{2012} B_y (1+r^*)^{1992-(y+1)} \right\} = \left\{ \sum_{y=1992}^{2012} C_y (1+r^*)^{1992-y} \right\}$$

Table 6.5 presents the evaluation metrics, for both lifecycle (1999–2022) and retrospective (1999–2012) streams of benefits and using both 7 percent and 3 percent discount rates for calculating NPV and BCR. Confidence intervals based on the variability of respondents' opinions are provided in parentheses after each metric, and additional explanation of their derivation is provided below.

The evaluation metrics in Table 6.5 suggest that VTO's investments in NiMH and Li-ion battery technologies have been socially valuable, even considering only the benefits through 2012. The most conservative estimates, ignoring benefits associated with the remaining useful life of attributable vehicles as of December 31, 2012, suggest that these investments were worthwhile as long as the IRR of the best alternative use of the funds would have been less than 12.8 percent, an assertion that is easy to defend.

Confidence intervals

Confidence intervals for the evaluation metrics reported in Table 6.5 were calculated based on our estimates of the market adoption of EDVs attributable to VTO as presented in Table 4.5 and Figure 4.6.[2] Recall that these estimates were obtained from interview and survey responses to Questions 5 and 7 from the interview guide. The confidence intervals around the metrics in Table 6.5 account for the variability of respondents' opinions based on 90 percent confidence intervals

Table 6.5 Evaluation metrics: economic and environmental health benefits, retrospective evaluation 1999 through 2012 (confidence intervals in parentheses)

Metric	*7 percent discount rate*	*3 percent discount rate*	*Internal rate of return*
Lifecycle evaluation, 1999–2022			
Net present value (base year =1992; millions of 2012$)	1,294 (1,025; 1,564)	3,334 (2,719; 3,949)	
Benefit-to-cost ratio	3.63 (3.08, 4.18)	5.74 (4.87, 6.61)	
Internal rate of return			17.7 percent (16.4 percent, 18.8 percent)
Retrospective evaluation, 1999–2012			
Net present value (base year =1992; millions of 2012$)	506 (361, 651)	1,303 (1,011; 1,595)	
Benefit-to-cost ratio	2.03 (1.73, 2.32)	2.85 (2.44, 3.27)	
Internal rate of return			14.3 percent (12.8 percent, 15.6 percent)

Note: Internal rates of return were derived using the combined benefits in column (1) of Table 6.4, by discounting these values back to January 1, 1992. This results in a small overestimation of the IRR because the discount rate applied in COBRA is fixed at 7 percent. The overestimation is small: a rough overcorrection still gives an IRR greater than 14 percent for retrospective and greater than 17 percent for lifecycle.

(plus or minus 1.65 standard deviations from the mean) around the percentage of market adoptions of EDVs attributable to VTO investments for each year, 1999–2012, reported in Table 4.5.

Recall that Table 4.6 translated the mean attributable percentage of market adoption for each year to a number of attributable EDVs, and that subsequent steps (illustrated with tables) translated these numbers of attributable EDVs to the metrics reported in Table 6.5. The confidence intervals for each metric in Table 6.5 were calculated by performing exactly the same steps – using, instead of the mean values in Table 4.5, the mean values plus or minus 1.65 their standard deviations.

Concluding observation

In the authors' judgment, it is appropriate to consider the benefits associated with the remaining useful life of attributable EDVs on the road on

December 31, 2012. We would therefore suggest that the most informative evaluation metrics are the BCR of 3.63 (based on a 7 percent discount rate) and the IRR of 17.7 percent. The confidence intervals on these metrics – (3.08, 4.18) for the BCR and (16.4 percent, 18.8 percent) for the IRR – reflect the variability of respondents' opinions.

Although these metrics are more optimistic than the strictly retrospective metrics that truncate the stream of benefits at December 31, 2012, one might still argue that they understate the net social benefits associated with VTP's R&D investments. We have noted before that the investments counted as costs included investments not strictly pertaining to NiMH or Li-ion technology. By limiting the scope of the study to benefits attributable to the accelerated adoption of EDVs embodying NiMH and Li-ion technology, we have implicitly assumed that the value of the VTO's investments in other technologies since 1992 has been zero, which is of course unlikely. We have also made no attempt to quantify the economic value of knowledge spillovers associated with VTO-funded R&D or other positive externalities, stemming for example from the retention of human capital in scientific fields related to energy storage technology.

A further issue is that we have included as costs all investments up to and including 2012 while including as benefits, even in the lifecycle analysis, only those associated with vehicles put on the road by December 31, 2012. A portion of VTO's R&D investments support the development of next-generation battery technologies, and as we learned from interviews it may take between 7 and 12 years for these technologies to be incorporated fully into EDVs and made commercially available. An example of this type of R&D would be the development of new anode materials for Li-ion batteries that will, as one respondent described, first appear in consumer electronics and may within ten years be scaled up for automotive application (as opposed to optimization of current cathodes for more immediate automotive application). Asked if the application of new Li-ion chemistries first to consumer electronics, almost as a proving ground prior to automotive application, was typical, this respondent said yes – or, if not typical, then certainly not unusual – citing as an example the use of lithium iron phosphate first for power tools. Many other Li-ion chemistries were first used in cell phones and laptops.

It is true that some VTO R&D investments could be expected to have shorter lags. As an example, funding from the American Recovery and Reinvestment Act (ARRA) was channeled through VTO to support the scale-up of production facilities for technologies that were already in place. In such cases, according to a senior R&D manager familiar with the program, the public investments would have had an observable impact within 2 to 3 years, allowing 18 months to 2 years to set up and 6 months to 1 year to validate the production lines. This 2- to 3-year lag could then reasonably be considered a lower bound.

As an illustrative example of the longer lag more typically associated with R&D investments, VTO supported the development of LG Chem Power pouch cell technology through successive USABC contracts over 10 years:

- DOE began evaluating the technology for EDV application with benchmark testing at Argonne National Laboratory (ANL) in 2002.
- USABC initiated a Technology Assessment Agreement with LG Chem Power in 2004.
- Following successful completion of that agreement, USABC entered into a technology development contract with LG Chem Power in 2006; a second phase development contract began in 2008.
- Following completion of the second phase development contract, GM began production of the Chevy Volt employing the USABC-developed LG Chem pouch cells in late 2011; Ford began production of the Focus EV, also employing the cells, in early 2012.

As this example illustrates, a portion of VTO investments of the past decade have supported technologies that have only recently had an economic impact. There is little doubt that it will be several more years before some technologies that received recent support from VTO's R&D investments will appear in vehicles on the road, as several other anecdotes make clear.

One respondent described the development of a new process for manufacturing cathodes and anodes that replaces the traditional solvent-based binder with a UV-curable binder. The process reduces curing time from 45 minutes to less than 1 second, replaces ovens the length of football fields with desk-sized UV lamps, and has the potential to reduce by half the cost of manufacturing EDV batteries. In this respondent's opinion, it is highly unlikely that this project would have been undertaken at all without VTO cost-sharing.

Another respondent described having recently completed a 3-year, $6 million project ($3 million in VTO funding) that developed a process to make a chemical that enables an increase in energy density of Li-ion cells: holding the size of the battery constant this process will result in a 30 percent increase in driving range, or, holding driving range constant the battery could be smaller and lighter.

Notes

1 See Chapter 1 and especially Note 22 for a discussion about the use of the 7 percent and 3 percent discount rates.

2 Regarding sensitivity analysis, OMB Circular A-94 (1992, p. 10) states: "Sensitivity Analysis: Major assumptions should be varied and net present value and other outcomes recomputed to determine how sensitive outcomes are to changes in the assumptions. The assumptions that deserve the most attention will depend on the dominant benefit and cost elements and the areas of greatest uncertainty of the program being analyzed. . . . In general, sensitivity analysis should be considered for estimates of: (i) benefits and costs; (ii) the discount rate; (iii) the general inflation rate; and (iv) distributional assumptions. Models used in the analysis should be well documented and, where possible, available to facilitate independent review."

7 Conclusions

Discussion of our findings

The economic evaluation summarized in this book demonstrates that VTO's R&D investments in NiMH and Li-ion battery technologies used in EDVs has been socially valuable. Gross benefits from fuel savings were quantified first for EDVs on the road between 1999 and 2012 ($3,433 million [2012$]), and second to account for these vehicles' remaining useful life from 2013 through 2022 ($4,217 million [2012$]) because these vehicles have an expected useful life of 11 years. The benefits calculated were only for VTO's contribution and therefore are 100 percent attributable to VTO.

When compared with VTO R&D investment costs of $971 million (2012$) from 1992 through 2012, the economic evaluation metrics for benefits through 2012 are:

- NPV of benefits: $506 million, discounted at 7 percent;
- BCR: 2.03-to-1, discounted at 7 percent; and
- IRR: 14.3 percent.

When those same investment costs are compared with the benefits estimated through 2022, for those vehicles on the road as of December 31, 2012, the evaluation metrics are:

- NPV of benefits: $1,294 million, discounted at 7 percent;
- BCR: 3.63-to-1, discounted at 7 percent; and
- IRR: 17.7 percent.

This second set of performance measures is more representative, because the continued operation of EDVs on the road as of December 2012 through the end of their effective useful lives is reasonably certain.

In addition, we find that:

- Over the period from 1999 through 2012, 44 percent of market adopted EDVs from 1992 through 2012 were, on average, attributable to VTO's R&D investments.

- The EDVs on the road from 1999 through 2012 are more fuel efficient than ICVs; gallons of gasoline saved per 1,000 attributable miles driven ranged from 13.8 for HEVs to 30.4 for EVs.
- Fuel savings benefits and environmental health benefits from 1999 through 2012 are attributable to VTO's R&D investments of $970.8 million in 2012 dollars from 1992 through 2012.
- The attributable fuel savings benefits from 1999 through 2012 totaled $3.3 billion in 2012 dollars, and from 1999 through 2022 they totaled $7.3 billion in 2012 dollars.
- The attributable environmental health benefits from 1999 through 2012 totaled $157 million, and from 1999 through 2022 they totaled $341 million.

Table 7.1 summarizes calculations from preceding sections, presenting measures and metrics that report the return on VTO's investments in energy storage along economic, energy, environmental, and energy security lines.

This evaluation's results are conservative. Although we conclude that VTO did indeed have a significant impact on NiMH technology, the impact on Li-ion is greater. However, only 3 years of Li-ion powered vehicles are included in this analysis (2010, 2011, and 2012), and evaluation participants predict the availability of new models in multiple vehicle segments and increasing adoption of Li-ion–powered vehicles over the next 5 years.

The following analytical points and assumptions support our conclusion that results are conservative, and these should be considered when interpreting and communicating evaluation results:

- Benefits for NiMH and Li-ion battery technology alone are compared with costs of VTO's entire energy storage R&D investment portfolio. Benefits were not considered for other technologies supported by VTO.
- Only vehicles on the road before January 1, 2013, were included in the benefit–cost analysis; vehicles entering operation on or after this date will certainly contain technology support by VTO.
- VTO's return on investment in Li-ion is expected to be greater than VTO's return on investment in NiMH, yet only 2 years of market adoption of Li-ion–powered vehicles are included in the benefits estimation.
- Industry experts and funded companies participating in initial interviews and our main data collection generally noted that the counterfactual technology development assessments they provided were lower-bound estimates. This is particularly so because it was not possible for them to measure the extent to which EDV models could have been redesigned in the face of inferior battery technical performance characteristics. Not only would some vehicles not have been on the road, but those that would have been on the road would likely have been less efficient.
- Newer cars and trucks may in actuality have effective useful lives longer than the average of 11 years documented in federal statistics.

Table 7.1 Summary benefit–cost analysis results

	Retrospective analysis through 2012	*Life-cycle analysis through 2022*	*Unit of measure*
Economic benefits and environmental health benefits			
Evaluation metrics			
Net present value @ 7% [Base year = 1992]	$506	$1,294	Million, 2012$
Net present value @ 3% [Base year = 1992]	$1,303	$3,334	Million, 2012$
Benefit-to-cost ratio @7%	2.03	3.63	
Benefit-to-cost ratio @3%	2.85	5.74	
Internal rate of return	14.3%	17.7%	
VTO R&D investments			
Present value @ 7% [Base year = 1992]	$492	$492	Million, 2012$
Present value @ 3% [Base year = 1992]	$703	$703	Million, 2012$
Economic benefits			
Present value @ 7% [Base year = 1992]	$952	$1,706	Million, 2012$
Present value @ 3% [Base year = 1992]	$1,904	$3,836	Million, 2012$
Environmental health benefits			
Monetized via COBRA			
Present value @ 7% [Base year = 1992]	$46	$81	Million, 2012$
Present value @ 3% [Base year = 1992]	$103	$202	Million, 2012$
Incidence			
Avoided mortality[a]	20.04	42.39	Deaths
Avoided infant mortality[a]	0.02	0.05	Deaths
Avoided nonfatal heart attacks	6.96	14.73	Attacks
Avoided resp. hospital admissions	4.34	9.17	Admissions
Avoided CDV hospital admissions	4.05	8.57	Admissions
Avoided acute bronchitis	18.24	38.57	Cases
Avoided upper respiratory symptoms	331.47	701.04	Episodes
Avoided lower respiratory symptoms	232.31	491.31	Episodes
Avoided asthma ER visits	7.79	16.48	Visits
Avoided MRAD	10,265.31	21,710.22	Incidences
Avoided work loss days	1,734.48	3,668.28	Days
Avoided asthma exacerbations	348.78	737.64	Episodes

(continued)

Table 7.1 (continued)

	Retrospective analysis through 2012	*Life-cycle analysis through 2022*	*Unit of measure*
Emissions benefits			
Avoided GHG emissions (CO_2eq)	6,989,237	14,461,042	Metric tons
Avoided volatile organic compounds emissions (VOCs)	3,928	7,926	Short tons
Avoided nitrogen oxides (NO_x)	1,217	2,324	Short tons
Avoided particulate matter emissions ($PM_{2.5}$)	2	16	Short tons
Avoided sulfur dioxide emissions (SO_2)	128	265	Short tons
Avoided ammonia emissions (NH_3)	643	1,329	Short tons
Energy and energy security benefits			
Avoided petroleum consumption	54,199,182	111,870,462	Barrels of oil
Avoided foreign petroleum consumption	27,226,445	47,833,782	Barrels of oil

Note

[a] Researchers have linked both short-term and long-term exposures to ambient levels of air pollution to increased risk of premature mortality. COBRA uses mortality risk estimates from an epidemiological study of the American Cancer Society cohort conducted by Krewski et al. (2009) and Laden et al. (2006). COBRA includes different mortality risk estimates for both adults and infants. Because of the high monetary value associated with prolonging life, mortality risk reduction is consistently the largest health endpoint valued in the study. The average of the low and high estimates of health benefits produced by COBRA was used for this study.

- EDVs' greater adoption in urban areas may have environmental benefits greater than what were estimated at the national level.
- There is a substantial lag between R&D investments and returns. Therefore, counting as costs all VTO investments through 2012 leads to conservative benefit–cost metrics when only the benefits associated with vehicles purchased through 2012 are counted.

The usual caveats of course apply. One should not generalize about the net benefits from EERE's R&D investments in other subprograms within VTO or within other energy areas. That said, one can reasonably ask how the findings presented here correspond to other economic evaluation conducted under the sponsorship of EERE, or for that matter, other economic evaluations that are conceptually comparable.

Comparable EERE evaluation studies

In earlier years EERE commissioned three retrospective evaluation studies. One study focused on geothermal technologies, a second on photovoltaic technologies, and a third on vehicle technologies. Each of these studies is comparable to the study summarized in this book in terms of methodology, although the technologies under study differed. In this section we summarize each of these studies in an effort to establish a benchmark for a relative comparison to EERE's investments in battery technology.[1]

Geothermal technologies

The purpose of this study was to estimate the social return on EERE's Geothermal Technologies Program's (GTP's) investments in geothermal technologies investment by comparing historical economic activity with GTP's current investment to the counterfactual of what would have likely happened in the absence of EERE's GTP. The study included an assessment of DOE's role in technology development and adoption, an estimate of the economic and environmental health benefits generated from selected technologies, and an estimate of measures of economic return from GTP's R&D activities.

Geothermal energy systems tap into thermal energy in the earth to produce heat and electricity. Geothermal power is a viable alternative for traditional fossil fuels (e.g., coal) or nuclear base-load generation. It also has the advantage of being a clean, renewable energy source without the variability of other renewable sources, such as wind and solar power. Resources of geothermal energy vary in quality and accessibility because of differences in depth of reservoirs, rock formations, and water content. Historically, geothermal power plants have been built under ideal conditions for energy production usually where the reservoir is close to the surface, the host rock is permeable and porous, and the ground fluid saturation and recharge rates allow economically feasible operation. The relative scarcity of high-quality natural geothermal sites has limited widespread geothermal energy use in the United States.

In the early 1970s, federally sponsored geothermal R&D began with funding from the Atomic Energy Commission and the National Science Foundation. The GTP was initiated by DOE in the late 1970s to support the development of technologies that would improve the economics of tapping geothermal resources. Since that time, GTP has conducted a wide range of research targeted at the long-term goal of making geothermal energy a cost-competitive power production alternative. This study selected four technologies that accrued significant economic benefits for the geothermal industry and other industries (e.g., oil and gas) for detailed analysis:

- Polycrystalline diamond compact (PDC) drill bits. PDC drill bits use harder, longer-lasting cutting surfaces and a simplified mechanical action; increasing both the productivity (more feet drilled per hour) and efficiency (less drill bits required per well) of drill bits.
- Binary cycle power plant technology. Binary cycle technology enabled the development of geothermal plants using low heat sources increasing geothermal capacity in place and offsetting electricity production by traditional base-load sources.

- TOUGH series of reservoir models. The TOUGH (Transport of Unsaturated Groundwater and Heat) series of models is a family of computer numerical simulation programs used to track fluid and heat flow in porous and fractured media. These models helped to optimize the performance of geothermal resources and manage risk associated with the uncertainty of their performance.
- High-temperature geothermal well cements. High-temperature geothermal well cements offer an improvement over alternative cement technology. They have a life expectancy of up to 20 years, eliminating annual reworks of geothermal and carbon dioxide injection wells. By comparison, wells that use traditional cement need to be reworked every one to two years.

For each of the four geothermal technologies selected, a common approach was used for the evaluation. This approach included the following five steps:

1 Conduct a historical review of the technology's development, demonstration, and commercialization (if applicable) to assess the R&D timeline and EERE's role.
2 Define the next best alternative technology.
3 Quantify the economic and environmental (air emissions) health net benefits by comparing the new (selected) technology to the next best alternative (independent of EERE attribution).
4 Determine the share of economic and environmental health net benefits attributable to DOE activities.
5 Calculate DOE program costs and estimate measures of economic performance.

The four technologies selected for analysis in this study reflected the wide range of research activities conducted by the GTP and were found, as a group, to have generated significant economic, environmental, and knowledge benefits.

- PDC drill bits. Approximately 60 percent of worldwide oil and gas well footage in 2006 was drilled using PDC drill bits (Blankenship, 2009). The main advantage of PDC drill bits over conventional roller cone bits is that they reduce the frequency of pulling the drill string to replace the drill bit, allowing higher penetration rates and thus reducing the time (and cost) of renting expensive drill rigs. The use of PDC drill bits in offshore applications in the oil and gas industry was estimated to reduce costs by $59 per foot drilled.
- Binary cycle. In reservoirs where the temperature range is 150°C to 190°C, flash cycle technology is economically viable but has approximately 15 percent lower electricity generation productivity as compared to binary cycle, because of its lower conversion efficiency. Thus, in this temperature range, the next best alternative is a traditional, but less productive, flash cycle geothermal plant.

- TOUGH models. Using reservoir modeling was estimated to have increased the productivity of geothermal resources by 10 percent. Although these benefits were somewhat offset by additional exploration costs associated with reservoir modeling, reservoir modeling was found to have been profitable in the aggregate for the geothermal industry by improving subsurface exploration.
- High-temperature cement. The rapid deterioration of Portland cement in geothermal wells (<12 months) required frequent and costly well remediation. The use of more structurally stable and corrosion-resistant high-temperature cements was estimated to eliminate $150,000 in annual well remediation costs and extend the working life of geothermal production wells to 20 years or more.

Streams of economic and environmental benefits attributable to DOE's investments in each of the four selected technologies were estimated based on interviews with industry and academic experts and compared with the total expenditures of GTP (including investments specific to these four technologies as well as others) to obtain the following evaluation metrics:

- NPV of benefits: $6.2 billion, using an annual discount rate of 7 percent;
- BCR: 4.7-to-1, also using a 7 percent discount rate; and
- IRR: 21 percent.

Photovoltaic technologies

The purpose of this study was to estimate the social return on EERE's Solar Energy Technology Program (SETP) investments in photovoltaic (PV) technology development. The technologies selected for analysis were PV module technologies. PV modules are encapsulated sets of solid-state solar cells that convert solar energy into electricity. They are perhaps most recognizable as the flat-plate solar panels mounted on roof tops, affixed to signal posts, or assembled in large arrays on solar farms. PV modules are usually characterized by the material technologies in the cells. These may be crystalline silicon (c-Si) or thin films of semiconductor material, particularly cadmium telluride (CdTe), copper indium diselenide (CIS), and amorphous silicon (a-Si).

PV technologies have benefited from long-term DOE investment that has supported core cell and module technology R&D, manufacturing process development, and the technology infrastructure supporting that R&D. Between 1975 and 2008, the period of analysis for this study, researchers in industry, academia, and DOE's national laboratories received financial and technical support to accelerate the development and market adoption of higher quality, longer lived, and lower cost PV modules.

The PV technologies reviewed in this report were developed with DOE funding or cost share under four initiatives. Each initiative represented a 10-year or longer commitment on the part of DOE to funding and technical expertise to researchers seeking to develop novel commercial technologies that exploit solar energy:

1 Flat-Plate Solar Array Project (FSA, 1975–1985), which was the first major terrestrial PV technology development initiative sponsored by the federal government. The project aggressively targeted core technical barriers so as to move photovoltaics from niche, off-grid applications to the mainstream. Technologies for silicon refining, encapsulants, automated module assembly, technology infrastructure, greater energy conversion efficiencies, silicon ingot growth, and silicon ribbon growth were developed. Industry experts interviewed for this study universally regarded the FSA period as foundational to the terrestrial PV industry.

In 1975, the U.S. PV industry produced 0.4 megawatts (MW) per year at an average cost per watt of $83.86 (2008$). Each module produced had no warranty and was expected to have a useful life of two to three years. When FSA ended in 1985, 7.8 MW was produced (a greater than 19-fold increase) at a production cost per watt of $9.40 (an 89 percent reduction), and 10-year warranties were offered.

2 PV Manufacturing Technology Project (PVMaT, 1991–2008), which targeted manufacturing operations to enable PV companies to accelerate decreases in production costs and increases in production capacity. PVMaT furthered low-cost PV module production via R&D into advanced manufacturing technologies for cell production and module assembly. Funded companies included AstroPower (GE), BP Solar, Evergreen, First Solar, Global Solar, SolarWorld USA, SunPower, and Uni-Solar.

In 1991, the U.S. PV industry produced 17.5 MW at a cost per watt of $6.93 (2008$). In 2008, 1,022.6 MW (a 58-fold increase) was produced at a cost per watt of $1.92 (a 72 percent reduction).

3 Thin-Film PV Partnerships (TFPs, 1994–2008), under which thin-film technologies were vastly improved, yielding thin-film PV modules that are produced today in greater numbers by U.S. manufacturers than c-Si modules. Through the 1980s and into the early1990s, NREL sponsored research that aimed to increase efficiency and reduce instability for a-Si, CdTe, and CIS PV technologies. U.S. PV companies reported receiving significant applied research funding beginning in 1988 under TFP's predecessor programs.

Thin films advanced dramatically during the past two decades, increasing from about 4 percent of all U.S. production in 1995 to over 60 percent in 2008. Steep production increases since 2005 were attributed to the success of major recipients of DOE funding funder TFP, including First Solar (CdTe), Global Solar (CIS/CIGS), and Uni-Solar (a-Si).

4 Measurement, characterization, and reliability R&D (1975–present), under which the technology infrastructure for module cell and reliability

(including the Outdoor Testing Facility), device performance, surface analysis, electro-optical characterization, and analytical microscopy was developed, provided an infrastructure that enabled industry, government, and university researchers to achieve their research objectives under the above three initiatives.

Technology infrastructure work for PV began in 1975 during FSA's block purchase program, which required NASA's Jet Propulsion Laboratory (JPL) and its contractors not only to design performance specifications but also to develop core measurement and characterization methods and standards for performance measurement. From that time to the time of the study, the infrastructure supporting the PV industry grew and became established, with private certifications, warranties, and Underwriters' Laboratories and International Electro-technical Commission standards.

This retrospective evaluation obtained the following evaluation metrics for the Photovoltaic Energy Systems Cluster:

- NPV of benefits: $1.5 billion, using a 7 percent discount rate;
- BCR: 1.8-to-1, using a 7 percent discount rate; and
- IRR: 17 percent.

The evaluation metrics obtained for the FSA were as follows, using a 7 percent discount rate for NPV and BCR:

- NPV of benefits: $2.4 billion;
- BCR: 7.1-to-1; and
- IRR: 37 percent.

For PVMaT (1991–2008) TFP (1988–2008) the evaluation metrics were as follows, using a 7 percent discount rate for NPV and BCR:

- NPV of benefits: $0.64 billion;
- BCR: 3.4-to-1; and
- IRR: 24 percent.

Vehicle technologies

The purpose of this study was to estimate the social return on EERE's Advanced Combustion Engine R&D (ACE R&D) sub-program within the Vehicle Technologies Program (VTP) Technology Program.

DOE began active R&D on vehicle technologies, with early emphasis on electric vehicle technology, as authorized by Congress through the Electric and Hybrid Vehicle Research, Development, and Demonstration Act (Public Law 94-413) of 1976. As part of the DOE mission, the Combustion Research Facility (CRF) at Sandia National Laboratories in Livermore, California, was formed in 1978 and began operations in 1981 with the intent of developing the most advanced possible diagnostic systems for combustion applications.

This study focused on two research areas funded by the ACE R&D sub-program: laser diagnostic and optical engine technologies, and combustion modeling. Both of these areas were relevant to heavy-duty diesel engines.

The selected technologies within these two research areas included:

- Laser Raman spectroscopy (LRS) uses a monochromatic light source (e.g., a laser) to probe a sample, and a detector measures the spectrum of frequencies contained in the light scattered in all directions from the sample. Molecules in the sample may either absorb radiation or contribute to the energy of scattered photons, resulting in a series of output frequencies that provide information about the molecules present.
- Laser Doppler velocimetry (LDV) measures the direction and speed of fluids (or other materials). Particle image velocimetry (PIV) is another diagnostic technique for measuring instantaneous velocity. Unlike LDV, PIV produces a two-dimensional vector field, while LDV measures velocity at a point.
- Mie scattering is an elastic scattering mechanism that occurs when light scatters off of particles with diameters on the same scale as the wavelength of light. The Mie scattering diagnostic is typically applied to particles in the 0.1–10 micron range. Fuel droplets exhibit Mie scattering when probed by lasers, and the scattered light can be collected by a detector to provide information about the spatial distribution of the droplets. Mie scattering is a useful phenomenon in a variety of combustion experiments, including those that focus on air flow and fuel spray. In-cylinder air flow can be observed and quantified in real time by scattering light off of particulates introduced into the air flow stream. Similarly, the distribution and evaporation of fuel droplets can be observed during diesel injection experiments. Information on spatial and temporal distribution is particularly useful for understanding and improving the dynamics of fuel injection.
- Rayleigh scattering is similar to Mie scattering, but it occurs with smaller particles and atoms or molecules in the gas phase. While Mie scattering occurs when the particle diameter is similar to the wavelength of incident light, Rayleigh scattering occurs when the particle diameter is much smaller than the wavelength of light.
- Laser-induced fluorescence (LIF) and tracer-based LIF are diagnostic tools that allow for the observation of light species such as the molecule OH and various molecular species that are common in combustion. Light species are particularly difficult to interrogate using other spectroscopic methods because very high energy (ultraviolet) sources are required for optical excitation. These species emit lower energy wavelengths that provide information about the vibrational-rotational states of molecules.
- Laser-induced incandescence (LII) is the emission of radiation that occurs when a laser beam interacts with soot or other particulate matter.

This technique can be used in the laboratory to determine information about average properties of soot that forms as a combustion product. The temperature of particulate matter rises when it absorbs incident laser light, and the heat generated is then emitted as thermal radiation. At very high temperatures, the soot or other particulate matter may vaporize.

- Improvements to semiconductor diode lasers that operate at room temperature in the visible and near-infrared areas of the spectrum have contributed to advances in the ways in which laser absorption spectrometry (LAS) is applied to combustion research. LAS is based on the principle that different molecular species absorb light of different wavelengths. New laser diodes have expanded the range of species that can be monitored using LAS; for example, lasers that emit in the infrared region have enabled better detection of species, such as carbon monoxide, that absorb infrared wavelengths. Improvements to sensor technologies that detect and identify the species present in a sample have also furthered the usefulness of LAS to combustion analysis. Because real-time monitoring is possible using LAS, the technique is employed to analyze engine combustion gas flows.
- Combustion modeling allows researchers to conduct "experiments" much more quickly than they could in the laboratory. Such modeling has thus expedited the discovery of new combustion engine technologies. The KIVA modeling software simulates the fluid dynamics of combustion processes in internal combustion engines.

The economic benefits considered relate to the reduced fuel consumption in heavy-duty diesel trucks resulting from research in and the application of laser and optical diagnostics and combustion modeling. The health and environmental benefits considered resulted from reduced diesel fuel consumption, which leads to reduced emissions, which in turn leads to reduced greenhouse gas and air pollutants.

This retrospective evaluation obtained the following evaluation metrics, using an annual discount rate of 7 percent for NPV and BCR:

- NPV of benefits: $23.1 billion;
- BCR: 153-to-1; and
- IRR: 63 percent.

Summary statement

Placing our battery study within the portfolio of other evaluation studies funded by EERE, the findings are similar although the quantitative values of the evaluation metrics differ. Differences in the quantitative values are due, in part, to the timing of each evaluation study. For example, recall from Table 3.3 that sales of HEVs using Li-ion battery technology began to grow in 2012, the last year considered for our evaluation. Thus, had the battery evaluation been conducted several years later, quantified benefits would have increased – perhaps dramatically – and thus each of the evaluation metrics would have increased in value. Still, regardless of time, it is clear that EERE's R&D investments in the technologies studied have been socially valuable.

Note

1 The following summaries draw on Gallaher, Link, and O'Connor (2012). More detailed descriptions of each study can be found there or in the respective reports: Gallaher et al. (2010) for geothermal, O'Connor et al. (2010) for photovoltaic, and Link (2010) for vehicle.

References

Anderson, D. L., and Patiño-Echeverri, S. (2009). *An Evaluation of Current and Future Costs for Lithium-ion Batteries for Use in Electrified Vehicle Powertrains*. Unpublished master's project. Durham, NC: Duke University.

Argonne National Laboratory (ANL). (2012, December). GREET life-cycle model. Retrieved from <http://greet.es.anl.gov/> (accessed 30 September 2014).

Arrow, K. J. (1962). Economic welfare and the allocation of resources for invention. In R. R. Nelson (Ed.) *The Rate and Direction of Inventive Activity: Economic and Social Factors*, Princeton, NJ: Princeton University Press, pp. 609–626.

Axsen, J., Burke, A., and Kurani, K. (2008). *Batteries for Plug-in Hybrid Electric Vehicles (PHEVs): Goals and the State of Technology circa 2008*. Davis, CA: University of California at Davis Institute of Transportation Studies.

Beresteanu, A., and Shanjun, L. (2011). Gasoline prices, government support, and the demand for hybrid vehicles in the U.S. *International Economic Review*, 52: 161–182.

Blackenship, D., Sandia National Laboratory (2009). Personal communication with Michael Gallaher and Alex Rogozhin, RTI International.

Bush, V. (1945). *Science – The Endless Frontier*, Washington, DC: National Science Foundation.

Dockery, D. W., Cunningham, J., Damokosh, A. I., Neas, L. M., Spengler, J. D., Koutrakis, P., et al. (1996). Health effects of acid aerosols on North American children: Respiratory symptoms. *Environmental Health Perspectives*, 104: 500–505.

Dooley, J. J. (2008). U.S. federal investments in energy R&D: 1961–2008. Pacific Northwest National Laboratory, U.S. Department of Energy.

Fehner, T., and Holl, J. M. (1994). Department of Energy 1977–1994: A summary history. Washington, DC: U.S. Department of Energy, History Division.

Gallaher, M. P., Link, A. N., and O'Connor, A. C. (2012). *Public Investments in Energy Technology*, London: Edward Elgar.

Gallaher, M. P., Rogozhin, A., and Petrusa, J. (2010). Retrospective benefit–cost evaluation of U.S. DOE geothermal technologies R&D program investments: Impacts of a cluster of energy technologies. Washington, DC: U.S. Department of Energy.

Griliches, Z. (1958). Research costs and social returns: Hybrid corn and related innovations. *Journal of Political Economy*, 66: 419–431.

Howell, D. (2011). Electric drive vehicle battery research activities at the U.S. Department of Energy. Paper presented at the 5th International Forum on Li-ion Battery Technology and Industrial Development, Beijing, China.

Hufschmidt, M. M. (2000). Benefit–cost analysis: 1922–1985. *Journal of Contemporary Water Research and Education*, 116: 42–49.

Huo, H., Wu, Y., and Wang, M. (2009). Total versus urban: Well-to-wheels assessment of criteria pollutant emissions from various vehicle/fuel systems. *Atmospheric Environment*, 43: 1796–1804.

Jaffe, A. B. (1998). The Importance of spillovers in the policy mission of the Advanced Technology Program. *The Journal of Technology Transfer*, 23: 11–19.

Jaffe, A. B., Newell, R. G., and Stavins, R. N. (2005). A tale of two market failures: Technology and environmental policy. *Ecological Economics*, 54: 164–174.

Krewski, D., Jerrett, M., Burnett, R. T., Ma, R., Hughes, E., Shi, Y., et al. (2009). Extended follow-up and spatial analysis of the American Cancer Society study linking particulate air pollution and mortality. *Research Report of the Health Effects Institute*, 140: 5–114; discussion 115–136.

Laden, F., Schwartz, J., Speizer, F. E., and Dockery, D. W. (2006). Reduction in fine particulate air pollution and mortality: Extended follow-up of the Harvard Six Cities Study. *American Journal of Respiratory and Critical Care Medicine*, 173: 667–672.

Link, A. N. (1999). Public/private partnerships in the United States. *Industry and Innovation*, 6: 191–217.

Link, A. N. (2006). *Public/Private Partnerships: Innovation Strategies and Policy Alternatives*, New York: Springer.

Link, A. N. (2010). Retrospective benefit–cost evaluation of U.S. DOE vehicle combustion engine R&D investments: Impacts of a cluster of energy technologies. Washington, DC: U.S. Department of Energy.

Link, A. N., and Link, J. R. (2009). *Government as Entrepreneur*, New York: Oxford University Press.

Link, A. N., and Scott, J. T. (1998). *Public Accountability: Evaluating Technology-Based Institutions*, Norwell, MA: Kluwer Academic Publishers.

Link, A. N., and Scott, J. T. (2010). *Public Goods, Public Gains: Calculating the Social Benefits of Public R&D*. New York: Oxford University Press.

Link, A. N., and Scott, J. T. (2012). The theory and practice of public-sector R&D economic impact analysis: Planning Report 11-1. Gaithersburg, MD: National Institute of Standards and Technology.

Lowe, M., Tokuoka, S., Trigg, T., and Gereffi, G. (2010). *Lithium-ion Batteries for Electric Vehicles: The U.S. Value Chain*. Duke University, Center on Globalization, Governance and Competitiveness, Durham, NC. Retrieved from <http://unstats.un.org/unsd/trade/s_geneva2011/refdocs/RDs/Lithium-Ion%20Batteries%20%28Gereffi%20-%20May%202010%29.pdf> (accessed 30 September 2014).

Mansfield, E., Rapoport, J., Romeo, A., Wagner, S., and Beardsley, G. (1977). Social and private rates of return from industrial innovations. *Quarterly Journal of Economics*, 91: 221–240.

Mar, T. F., Larson, T. V., Stier, R. A., Claiborn, C., and Koenig, J. Q. (2004). An analysis of the association between respiratory symptoms in subjects with asthma and daily air pollution in Spokane, Washington. *Inhalation Toxicology*, 16: 809–815.

Martin, S., and Scott, J. T. (2000). The nature of innovation market failure and the design of public support for private innovation. *Research Policy*, 29: 437–447.

Moolgavkar, S. H. (2000). Air pollution and hospital admissions for chronic obstructive pulmonary disease in three metropolitan areas in the United States. *Inhalation Toxicology*, 12(Supplement 4): 75–90.

Moolgavkar, S. H. (2003). Air pollution and daily deaths and hospital admissions in Los Angeles and Cook Counties. In *Revised Analyses of Time-Series Studies of Air Pollution and Health*, Boston, MA: Health Effects Institute, pp. 183–198.

National Institute of Standards and Technology (NIST). (1997). Economic assessment of the NIST thermocouple calibration program: Planning Report 97-1. Gaithersburg, MD: National Institute of Standards and Technology.

National Institute of Standards and Technology (NIST). (1999). Economic assessment of the NIST alternative refrigerants research program: Planning Report 98-1. Gaithersburg, MD: National Institute of Standards and Technology.

National Institute of Standards and Technology (NIST). (2000). The economic impact of NIST's cholesterol standards program: Planning Report 00-4. Gaithersburg, MD: National Institute of Standards and Technology.

National Research Council (NRC). (1998). *Effectiveness of the United States Advanced Battery Consortium as a Government-Industry Partnership*. Washington, DC: National Academy Press.

National Research Council (NRC). (2001). *Energy Research at DOE: Was It Worth It? Energy Efficiency and Fossil Energy Research 1978 to 2000*. Washington, DC: National Academy Press.

Nixon, R. M. (1974). Address on the state of the union delivered before a joint session of the Congress, January 30, 1974. Retrieved from <www.presidency.ucsb.edu/ws/index.php?pid=4327> (accessed 30 September 2014).

Novák, P., Goers, D., and Spahr, M. E. (2010). Carbon materials in Lithium-ion batteries. In F. Béguin and E. Frackowiak (Eds.) *Carbons for Electrochemical Energy Storage and Conversion Systems*, London: CRC Press, pp. 263–328.

Oakes, M. J. (2006). A brief history of batteries and stored energy, *Neta World* Summer: 1–6.

O'Connor, A. C., Loomis, R. J., and Braun, F. M. (2010). Retrospective benefit–cost evaluation of DOE investment in photovoltaic energy systems. Washington, DC: U.S. Department of Energy.

O'Connor, A. C., Gallaher, M. P., Loomis, R. J., and Casey, S. E. (2013). Estimating avoided environmental emissions and environmental health benefits. In A. N. Link and N. S. Vonortas (Eds.) *Handbook on the Theory and Practice of Program Evaluation*, Cheltenham: Edward Elgar, pp. 247–290.

Office of Management and Budget (OMB). (1992). Circular No. A–94: Guidelines and discount rates for benefit–cost analysis of federal programs. Washington, DC: U.S. Government Printing Office.

Office of Management and Budget (OMB). (2003). Circular No. A–4: Regulatory analysis, Washington, DC: U.S. Government Printing Office.

Office of Technology Assessment (OTA). (1995). Advanced automotive technology: Visions of a super-efficient family car. OTA-ETI-638, Washington, DC: U.S. Government Printing Office.

Office of Technology Policy. (1996). Effective Partnering: A Report to Congress on Federal Technology Partnerships, Washington, DC: U.S. Department of Commerce.

Ostro, B. D. (1987). Air pollution and morbidity revisited: A specification test. *Journal of Environmental Economics and Management*, 14: 87–98.

Ostro, B. D., Lipsett, M., Mann, J., Braxton-Owens, H., and White, M. (2001). Air pollution and exacerbation of asthma in African-American children in Los Angeles. *Epidemiology*, 12: 200–208.

Ostro, B. D., and Rothschild, S. (1989). Air pollution and acute respiratory morbidity: An observational study of multiple pollutants. *Environmental Research*, 50: 238–247.

Peters, A., Dockery, D. W., Muller, J. E., and Mittleman, M. A. (2001). Increased particulate air pollution and the triggering of myocardial infarction. *Circulation*, 103: 2810–2815.

Pope, C. A., Burnett, R. T., Thun, M. J., Calle, E. E., Krewski, D., Ito, K., and Thurston, G. D. (2002). Lung cancer, cardiopulmonary mortality, and long-term exposure to fine particulate air pollution. *Journal of the American Medical Association*, 287: 1132–1141.

RTI International. (2013). Benefit–cost evaluation of U.S. DOE investment in energy storage technologies for hybrid and electric cars and trucks. Final report to the Office of Energy Efficiency and Renewable Energy, U.S. Department of Energy.

Ruegg, R. R., and Jordan, G. (2011). Guide for conducting benefit–cost evaluation of realized impacts of public R&D programs, Washington, DC: U.S. Department of Energy.

Ruegg, R. R., and Thomas, P. (in press). Knowledge benefits from DOE-funded research in energy storage R&D for hybrid and electric vehicles. Washington, DC: U.S. Department of Energy.

Schwartz, J., and Neas, L. M. (2000). Fine particles are more strongly associated than coarse particles with acute respiratory health effects in schoolchildren. *Epidemiology*, 11: 6–10.

Sheppard, L. (2003). Ambient air pollution and nonelderly asthma hospital admissions in Seattle, Washington, 1987–1994. In *Revised Analyses of Time-Series Studies of Air Pollution and Health*, Boston, MA: Health Effects Institute, pp. 227–230.

Sheppard, L., Levy, D., Norris, G., Larson, T. V., and Koenig, J. Q. (1999). Effects of ambient air pollution on nonelderly asthma hospital admissions in Seattle, Washington, 1987–1994. *Epidemiology*, 10: 23–30.

Sissine, F. (1996). *The Partnership for a New Generation of Vehicles (PNGV)*. CRS Report 96-191. Retrieved from <http://ftp.ncseonline.org/nle/crs/abstract.cfm?NLEid=801> (accessed 30 September 2014).

Sperling, D. (2001). Public-private technology R&D partnerships: Lessons from U.S. partnership for a new generation of vehicles. *Transport Policy*, 8: 247–256.

Tassey, G. (2003). *Methods for Assessing the Economic Impacts of Government R&D: NIST Planning Report 03-1*. Gaithersburg, MD: National Institute of Standards and Technology.

Trinkle, D. S. (2009). A Vehicle for Change: PNGV, an Experiment in Government-Industry Cooperation. The Pardee RAND Graduate School.

U.S. Census Bureau. (annual). *Statistical Abstract of the United States*. Retrieved from <www.census.gov/prod/2011pubs/12statab/pop.pdf> (accessed 30 September 2014).

U.S. Department of Commerce, Bureau of Economic Analysis, downloaded from the St. Louis Federal Reserve, 3–4. Retreived from <http://research.stlouisfed.org/fred2/series/GDPCTPI/downloaddata?cid=21> (accessed 30 September 2014).

U.S. Department of Energy (DOE). (2012). Transportation energy data book, edition 31. Washington, DC: U.S. Department of Energy.

U.S. Department of Energy (DOE). (2013). Ethanol blends. Retrieved from <www.afdc.energy.gov/fuels/ethanol_blends.html> (accessed 30 September 2014).

U.S. Energy Information Administration (EIA). (2012a). AEO2013: Figure 11—U.S. liquid fuels supply, 1970-2040 (million barrels per day). Retrieved from <www.eia.gov/forecasts/aeo/er/excel/overview.fig11.data.xls> (accessed 30 September 2014).

U.S. Energy Information Administration (EIA). (2012b). Frequently asked questions: How many gallons of gasoline does one barrel of oil make? Retrieved from <www.eia.gov/tools/faqs/faq.cfm?id=24andt=10> (accessed 30 September 2014).

U.S. Energy Information Administration (EIA) (2013). Annual Energy Outlook 2013, Washington, DC, U.S. Department of Energy.

U.S. Environmental Protection Agency (U.S. EPA). (2000). Control of air pollution from new motor vehicles: Tier 2 motor vehicle emissions standards and gasoline sulfur control requirements. *Federal Register*, 65: 6698–6870.

U.S. Environmental Protection Agency (U.S. EPA). (2012a). Greenhouse gas emissions. Retrieved from <www.epa.gov/climatechange/ghgemissions/gases.html> (accessed 30 September 2014).

U.S. Environmental Protection Agency (U.S. EPA). (2012b). Greenhouse gas equivalencies calculator. Retrieved from <www.epa.gov/cleanenergy/energy-resources/calculator.html> (accessed 30 September 2014).

U.S. Environmental Protection Agency (U.S. EPA). (2012c). The social cost of carbon. Retrieved from <www.epa.gov/climatechange/EPAactivities/economics/scc.html> (accessed 30 September 2014).

U.S. Environmental Protection Agency (U.S. EPA). (2012d). User's manual for the Co-Benefits Risk Assessment (COBRA) screening model. Retrieved from <www.epa.gov/statelocalclimate/documents/pdf/COBRA_manual.pdf> (accessed 30 September 2014).

U.S. Environmental Protection Agency (U.S. EPA). (2012e). What are the six common air pollutants? Retrieved from <www.epa.gov/air/urbanair/> (accessed 30 September 2014).

U.S. Environmental Protection Agency (U.S. EPA). (2013). MOVES (motor vehicle emission simulator). Retrieved from <www.epa.gov/otaq/models/moves/index.htm> (accessed 30 September 2014).

Woodruff, T. J., Grillo, J., and Schoendorf, K. C. (1997). The relationship between selected causes of postneonatal infant mortality and particulate air pollution in the United States. *Environmental Health Perspectives*, 105(6): 608–612.

Woods & Poole Economics Inc. (2011). 2012 complete U.S. demographic database files. Washington, DC: Woods & Poole Economics Inc.

Wu, Y., Wang, M. Q., Sharer, P. B., and Rousseau, A. (2006). Well-to-wheels results of energy use, greenhouse gas emissions and criteria air pollutant emissions of selected vehicle/fuel systems. 2006-01-0377. New York: SAE International. Retrieved from <www.autonomie.net/docs/6%20-%20Papers/WTW/wtw_selected_vehicle_fuels_systems.pdf> (accessed 30 September 2014).

Index